Masoud Soroush (Ed.)

Polymer Reactor Modeling, Design and Monitoring

MDPI

This book is a reprint of the Special Issue that appeared in the online, open access journal, *Processes* (ISSN 2227-9717) from 2015–2016, available at:

http://www.mdpi.com/journal/processes/special_issues/polymer_modeling

Guest Editor
Masoud Soroush
Department of Chemical and Biological Engineering,
Drexel University
USA

Editorial Office
MDPI AG
St. Alban-Anlage 66
Basel, Switzerland

Publisher
Shu-Kun Lin

Assistant Editor
Jennifer Li

1. Edition 2016

MDPI • Basel • Beijing • Wuhan • Barcelona • Belgrade

ISBN 978-3-03842-254-9 (Hbk)
ISBN 978-3-03842-255-6 (electronic)

Table of Contents

List of Contributors

Ahmad Arabi Shamsabadi Department of Chemical and Biological Engineering, Drexel University, Philadelphia, PA 19104, USA.

Kathryn A. Berchtold Carbon Capture and Separations for Energy Applications, Materials Physics and Applications Division, Los Alamos National Laboratory, Los Alamos, NM 87545, USA.

Carlos A. Castor Jr. Department of Physics and Engineering Physics, Tulane University, New Orleans, LA 70118, USA.

Patrick Corcoran Department of Chemical and Biological Engineering, Drexel University, Philadelphia, PA 19104, USA.

Michael F. Drenski Advanced Polymer Monitoring Technologies, Inc., New Orleans, LA 70125, USA.

Marc A. Dubé Department of Chemical and Biological Engineering, Centre for Catalysis Research and Innovation, University of Ottawa, 161 Louis Pasteur Pvt., Ottawa, ON K1N 6N5, Canada.

Aryan Geraili Department of Chemical Engineering, Louisiana State University, Baton Rouge, LA 70803, USA.

Navid Ghadipasha Department of Chemical Engineering, Louisiana State University, Baton Rouge, LA 70803, USA.

Michael C. Grady DuPont Experimental Station, Wilmington, DE 19803, USA.

Biao Huang Department of Chemical and Materials Engineering, University of Alberta, 12th Floor—Donadeo Innovation Centre for Engineering (ICE), 9211—116 Street, Edmonton, AB T6G 1H9, Canada.

Costas Kravaris Artie McFerrin Department of Chemical Engineering, Texas A & M University, College Station, TX 77843, USA.

Ruoxia Li Department of Chemical and Materials Engineering, University of Alberta, 12th Floor—Donadeo Innovation Centre for Engineering (ICE), 9211—116 Street, Edmonton, AB T6G 1H9, Canada.

Fernando V. Lima Department of Chemical Engineering, West Virginia University, Morgantown, WV 26506, USA.

Chen Ling Artie McFerrin Department of Chemical Engineering, Texas A & M University, College Station, TX 77843, USA.

Chandra Mouli R. Madhuranthakam Department of Chemical Engineering, 200 University Avenue West, Waterloo, ON N2L 3G1, Canada.

Nazanin Moghadam Department of Chemical and Biological Engineering, Drexel University, Philadelphia, PA 19104, USA.

Alexander Penlidis Institute for Polymer Research (IPR), Department of Chemical Engineering, 200 University Avenue West, Waterloo, ON N2L 3G1, Canada; Institute for Polymer Research (IPR), Department of Chemical Engineering, University of Waterloo, Waterloo, ON N2L 3G1, Canada.

Vinay Prasad Department of Chemical and Materials Engineering, University of Alberta, 12th Floor—Donadeo Innovation Centre for Engineering (ICE), 9211—116 Street, Edmonton, AB T6G 1H9, Canada.

Andrew J. Radcliffe Department of Chemical Engineering, West Virginia University, Morgantown, WV 26506, USA.

Andrew M. Rappe The Makineni Theoretical Laboratories, Department of Chemistry, University of Pennsylvania, Philadelphia, PA 19104-6323, USA.

Wayne F. Reed Department of Physics and Engineering Physics, Tulane University, New Orleans, LA 70118, USA.

Shanshan Ren Department of Chemical and Biological Engineering, Centre for Catalysis Research and Innovation, University of Ottawa, 161 Louis Pasteur Pvt., Ottawa, ON K1N 6N5, Canada.

Marzieh Riahinezhad Institute for Polymer Research (IPR), Department of Chemical Engineering, University of Waterloo, Waterloo, ON N2L 3G1, Canada.

Jose A. Romagnoli Department of Chemical Engineering, Louisiana State University, Baton Rouge, LA 70803, USA.

Alison J. Scott Institute for Polymer Research (IPR), Department of Chemical Engineering, University of Waterloo, Waterloo, ON N2L 3G1, Canada.

Rajinder P. Singh Carbon Capture and Separations for Energy Applications, Materials Physics and Applications Division, Los Alamos National Laboratory, Los Alamos, NM 87545, USA.

Masoud Soroush Department of Chemical and Biological Engineering, Drexel University, Philadelphia, PA 19104, USA.

Sriraj Srinivasan Arkema Inc., 900 1st Avenue, King of Prussia, PA 19406, USA.

Hidetaka Tobita Department of Materials Science and Engineering, University of Fukui, Fukui 910-8507, Japan.

Eduardo Vivaldo-Lima Facultad de Química, Departamento de Ingeniería Química, Universidad Nacional Autónoma de México, México D.F. 04510, Mexico.

About the Guest Editor

Masoud Soroush is a Professor of Chemical and Biological Engineering at Drexel University, Philadelphia, PA, U.S.A. He received his B.S. in Chemical Engineering from Abadan Institute of Technology, Iran, in 1985, and his M.S. in Chemical Engineering in 1988, M.S. in Electrical Engineering: Systems in 1991, and Ph.D. in Chemical Engineering in 1992, all from the University of Michigan, Ann Arbor, MI. He was a Visiting Scientist at DuPont Marshall Lab, Philadelphia, 2002–2003, and a Visiting Professor at Princeton University, Princeton, NJ, in 2008. His research interests are in polymer engineering; quantum chemical calculations; process systems engineering; probabilistic modeling, inference and rare-event prediction; and mathematical modeling, analysis and optimization of renewable power generation and storage systems. He has authored or co-authored over 160 refereed papers. His awards include a U.S. National Science Foundation Faculty Early CAREER Award in 1997 and an American Automatic Control Council O. Hugo Schuck Best Paper Award in 1999. He is a fellow of the American Institute of Chemical Engineers.

Preface to "Polymer Reactor Modeling, Design and Monitoring"

Polymers range from synthetic plastics, such as polyacrylates, to natural biopolymers, such as proteins and DNA. The large molecular mass of polymers and our ability to manipulate their compositions and molecular structures have allowed for producing synthetic polymers with attractive properties. As such, synthetic polymers have been increasingly used in a large number of applications such as paints, coatings, fibers, flexible films, automotive parts, adhesives, fuel cells, batteries, medicine, and controlled drug delivery. For example, only in 2013, around 299 million tons of plastics were produced, and this level has increased since then. Worldwide polymer production is expected to grow; as polymers steadily replace materials such as glass, wood and metals, our understanding of polymers improves, and new polymers with remarkable characteristics are synthesized. Because of the huge production volume of commodity polymers, a little improvement in the operation of commodity-polymer processes can lead to significant economic gains. On the other hand, a little improvement in the quality of specialty polymers can lead to substantial increase in economic profits.

This Special Issue includes papers that investigate different approaches to improving polymers, polymer processes, and processes that use polymers. These approaches include state and parameter estimation in polymerization reactors, polymerization reactor modeling for process monitoring, polymerization reactor monitoring, design of optimal polymerization experiments, use of polymeric membranes in integrated gasification combined cycle units, model-based design of polymerization reactors, and polymerization reactor modeling.

Masoud Soroush
Guest Editor

Model-Based Reactor Design in Free-Radical Polymerization with Simultaneous Long-Chain Branching and Scission

Hidetaka Tobita

Abstract: Polymers are the products of processes and their microstructure can be changed significantly by the reactor systems employed, especially for nonlinear polymers. The Monte Carlo simulation technique, based on the random sampling technique, is used to explore the effect of reactor types on the branched polymer structure, formed through free-radical polymerization with simultaneous long-chain branching and scission, as in the case of low-density polyethylene synthesis. As a simplified model for a tower-type multi-zone reactor, a series of continuous stirred-tank reactors, consisting of one big tank and the same N-1 small tanks is considered theoretically. By simply changing the tank arrangement, various types of branched polymers, from star-like globular structure to a more randomly branched structure, can be obtained, while keeping the following properties of the final products, the monomer conversion to polymer, the average branching and scission densities, and the relationship between the mean-square radius of gyration and molecular weight.

Reprinted from *Processes*. Cite as: Tobita, H. Model-Based Reactor Design in Free-Radical Polymerization with Simultaneous Long-Chain Branching and Scission. *Processes* **2015**, *3*, 731–748.

1. Introduction

Nonlinear polymer formation under a kinetically controlled condition is, in general, history-dependent, and the history of every polymer molecule determines the properties of final product polymers. The molecular architecture can be very complex; however, the history-dependence opens up the opportunities to control the nonlinear structure through various types of reactor operation.

In this article, free-radical polymerization that involves chain transfer to polymer, leading to long-chain branching and scission, as in the case of high-pressure polymerization of ethylene to produce low-density polyethylene [1,2], is considered. As shown in Figure 1, the birth time of the chains, C_1 and C_2 must be some time after that of A or A'. There is a definite time order for the chain connection statistics.

When the scission reaction is involved together with long-chain branching, the time sequence of branching and scission must be properly accounted for. This kind of reaction system cannot be fully represented by a simple set of population balance

1

differential equations [3–7]. On the other hand, by application of Monte Carlo (MC) method, based on the random sampling technique [8,9], history-dependence of branching and scission can be fully accounted for [7,10–12]. In this MC simulation method, the structure of each polymer molecule can be observed directly on the computer screen, and very detailed structural information can be obtained. On the basis of such detailed structural information, it is possible to determine the viscoelastic properties of branched polymers [13].

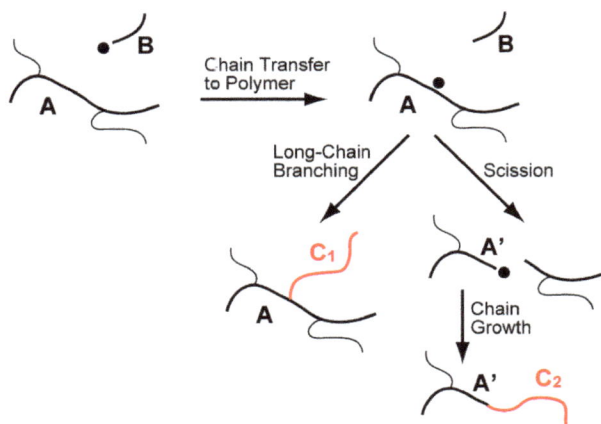

Figure 1. Schematic representation of the process of chain transfer to polymer, leading to long-chain branching and scission.

This simulation method can be used to investigate the effect of reactor type on the formed branched structure. Following the fundamentals of chemical reaction engineering [14], consider how the difference of an ideal plug flow reactor (PFR) and an ideal continuous stirred-tank reactor (CSTR) affects the present polymerization system. Note that a PFR is equivalent to a batch reactor, in which the length in a PFR can be converted to time in a batch reactor. Figure 2 shows the MC simulation results [12] of the weight fraction distribution for the polymers synthesized in a PFR and in a CSTR. In the simulation, the final average branching density ($\bar{\rho} = 2.22 \times 10^{-3}$), as well as the final scission density ($\bar{\eta} = 1.11 \times 10^{-4}$), is set to be the same for both types of reactors. The independent variable shown in the figure is the logarithm of degree of polymerization (DP), $\log_{10} P$, as usually employed in the gel permeation chromatography (GPC) measurement. The high DP tail extends more significantly in a CSTR, and the weight-average DP is larger for the product in a CSTR, even though the average branching density level is set to be the same.

Figure 3 shows the MC simulation results [12] for the relationship between the branching density ρ and the degree of polymerization P. The branching density is defined as the fraction of units having a branch point. In Figure 3a,b, each dot represents ρ and P of each polymer molecule simulated, and the solid blue curve with circular symbols shows the average ρ within small ΔP intervals, which is the estimate of the average branching density of polymers having degree of polymerization P, $\bar{\rho}(P)$. The dashed black line shows the average branching density of the whole system, $\bar{\rho}$. The value of $\bar{\rho}(P)$ increases with P, but reaches a constant limiting value $\rho_{P \to \infty}$, and $\rho_{P \to \infty}$ is larger than the average branching density of the whole system, $\bar{\rho}$. This is a rather general characteristic of the branched polymer systems [9], including the hyperbranched polymers [15]. Figure 3c shows the comparison of $\bar{\rho}(P)$ for a PFR and for a CSTR. The limiting branching density is larger for the CSTR.

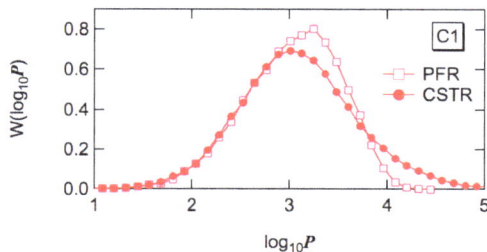

Figure 2. Weight fraction distribution of polymers formed in a PFR and in a CSTR, when the average branching density, as well as the average scission density, is set to be the same for both types of reactors [12]. The kinetic parameters used are the same as C1, shown later in Table 2 of this article.

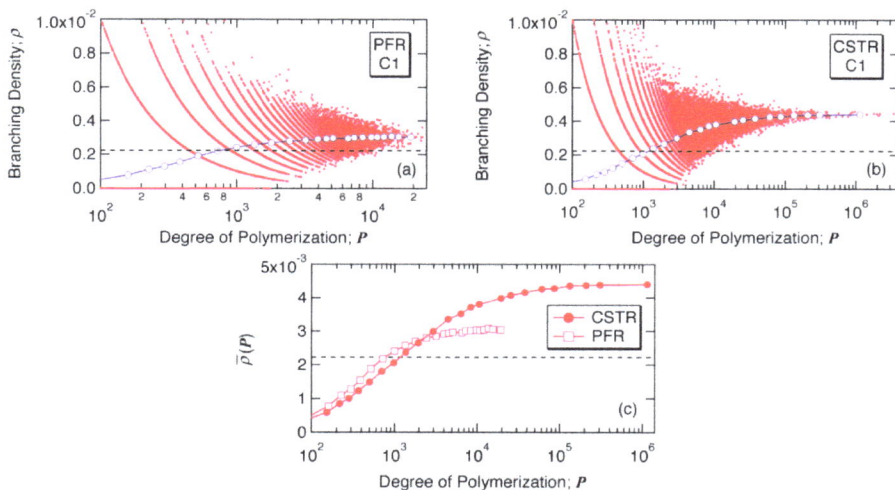

Figure 3. Relationship between branching density ρ and degree of polymerization P for the branched polymers formed in (**a**) a PFR and in (**b**) a CSTR, and (**c**) $\bar{\rho}(P)$ for both types of reactors. The calculation condition is the same as in Figure 2.

3

Figure 4 shows the relationship between the mean square radius of gyration for the unperturbed chains $<s^2>_0$ and P. In the figure, b is a constant defined by $b = u/l^2$, where u is the number of monomeric units in a random walk segment, and l is the length of a random walk segment. Each dot in Figure 4 represents $<s^2>_0$ and P-value of each polymer molecule simulated, and the circular symbols show the average values of $<s^2>_0$ within small ΔP intervals, which are the estimates of the average $<s^2>_0$ of polymers having DP, P. For randomly branched polymers, the radii of gyration is known to be given by the Zimm-Stockmayer equation [16], represented by:

$$b \left\langle s^2 \right\rangle_0 = \frac{F}{6} \left[\left(1 + \frac{m_P}{7} \right)^{0.5} + \frac{4 m_P}{9 \pi} \right]^{-0.5} \tag{1}$$

where m_P is the number of branch points with degree of polymerization P, and for the present case it can be estimated from:

$$m_P = P \overline{\rho}(P) \tag{2}$$

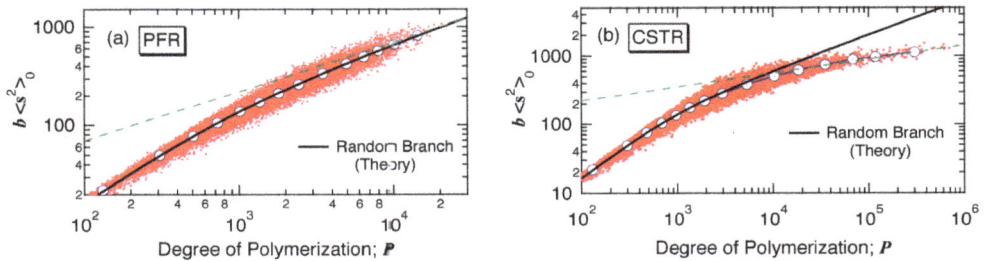

Figure 4. Relationship between the mean-square radius of gyration $<s^2>_0$ and degree of polymerization P for the branched polymers formed in (**a**) a PFR and in (**b**) a CSTR [12]. In the figure, b is a constant defined by $b = u/l^2$, where u is the number of monomeric units in a random walk segment, and l is the length of a random walk segment. The black solid curve shows the relationship for random branched polymers. The slope of broken straight line (green) for (**a**) is 0.5, while that for (**b**) is 0.2.

In Figure 4, the radii of gyration of randomly branched polymers having the branching density, $\overline{\rho}(P)$ is shown by a black curve. For a PFR, the radius of gyration is essentially the same as for the random branched polymers, as shown in Figure 4a. In a PFR or a batch reactor, $<s^2>_0$ is close to that for the random branched polymers, at least for low conversion regions, while the $<s^2>_0$-values tend to become slightly larger than those for the random branched polymers at higher conversions [17].

4

On the other hand, for a CSTR, the $<s^2>_0$-values for large polymers are significantly smaller than the random branched polymers, as shown in Figure 4b.

Incidentally, because the $\bar{\rho}(P)$-value reaches a constant value at large P's, the $<s^2>_0$ curve for the random branched polymer, represented by Equation (1), follows the power law with $<s^2>_0 \propto P^{0.5}$ for $P \rightarrow \infty$. In Figure 4a, the broken line shows the slope with 0.5. On the other hand, for a CSTR, it is not clear if the power law holds for large polymers, but the broken line drawn as a trial shows the slope with 0.2.

Table 1 summarizes the characteristics of PFR and CSTR for free-radical polymerization with chain transfer to polymer. Note that Table 1 applies for low scission frequency cases. Because the polymer transfer reaction is the reaction between a polymer and a polymer radical, larger polymer concentration promotes the chain transfer reaction. With a CSTR, the polymer concentration is high throughout the polymerization. If the final conversion is set to be the same for both types of reactors, the average branching density is larger for the polymers synthesized in a CSTR. In the present example, because the final average branching density is set to be the same, the final conversion must set to be larger for a PFR. It is interesting to note that the weight-average DP, $\bar{P}_w$ is larger for a CSTR, even though the average branching and scission densities are the same for both types of reactors. As shown in Figure 2, the molecular weight distribution of polymers formed in a CSTR is broader.

Table 1. Comparison of PFR and CSTR, under condition where the final average branching density is set to be the same for both types of reactors.

Property of product polymer	PFR (or Batch)	CSTR
Final conversion	$>$	
Weight-average DP, $\bar{P}_w$	$<$	
Branching density of large polymers, $\rho_{P \rightarrow \infty}$	$<$	
$<s^2>_0$, compared with random branched polymers	Almost the same. (May become larger at higher conversions.)	Significantly smaller for large polymers.

The branching density of large polymers, represented by $\rho_{P \rightarrow \infty}$ is larger for a CSTR, as shown in Figure 3. For the polymers synthesized in a PFR, the radius of gyration of polymers having the degree of polymerization P is essentially the same as that for the random branched polymers. It would be reasonable to consider that the branched structure formed in a PFR is close to randomly branched. On the other hand, as Figure 4b shows, the polymers formed in a CSTR possess much smaller radius of gyration compared with that for the random branched polymers with the same P-value.

5

As the textbook describes [14], when the multiple CSTRs are connected in series, the residence time distribution approaches to that of a PFR, as the number of tanks increases. The tanks-in-series model was applied to the present reaction system [18] to confirm that as the number of tanks increases the produced polymers approach to those formed in a PFR. Figure 5 shows how the relationship between $<s^2>_0$ and P changes by increasing the total number of tanks.

The MC simulation method employed in refs. [7,12,18] can be used to investigate the effect of reaction environment change on the formed branched polymer structure, which may lead to a discovery of novel type of reactor systems. In this article, a series of continuous stirred-tank reactors, consisting of one big tank and the same N-1 small tanks is discussed by using the data reported in reference [7] and newly created MC simulation data. The possibility of model-based reactor design for the nonlinear polymerization system is explored.

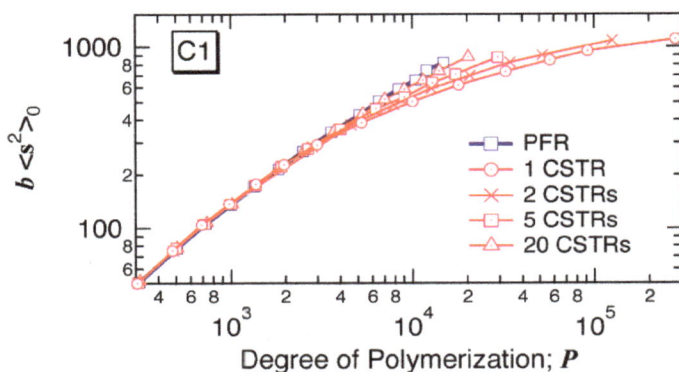

Figure 5. Effect of the number of CSTRs in the tanks-in-series model on the relationship between the mean-square radius of gyration $<s^2>_0$ and degree of polymerization P. The MC simulation data were taken from [18].

2. Simulation Method

The random sampling technique [8,9] is used to estimate the properties of final product polymers. In this method, a polymer molecule is selected randomly from the product polymers, and the molecular structure is reconstructed by following the history of this particular polymer molecule, as schematically represented by Figure 6.

When the random sampling technique is applied to the Monte Carlo method, a large number of polymers are sampled, and the statistical properties of the whole system are determined effectively. In this method, the system size considered is infinitely large, and therefore, the system boundary problem does not occur. In addition, the amount of calculation required does not increase significantly by increasing the number of CSTRs in series. This MC simulation does not proceed in the order of reaction time, as in the most of other MC methods [19]. Looking from the given chain, the chains with different birth times are connected back and forth by following the reaction history that particular polymer molecule has experienced.

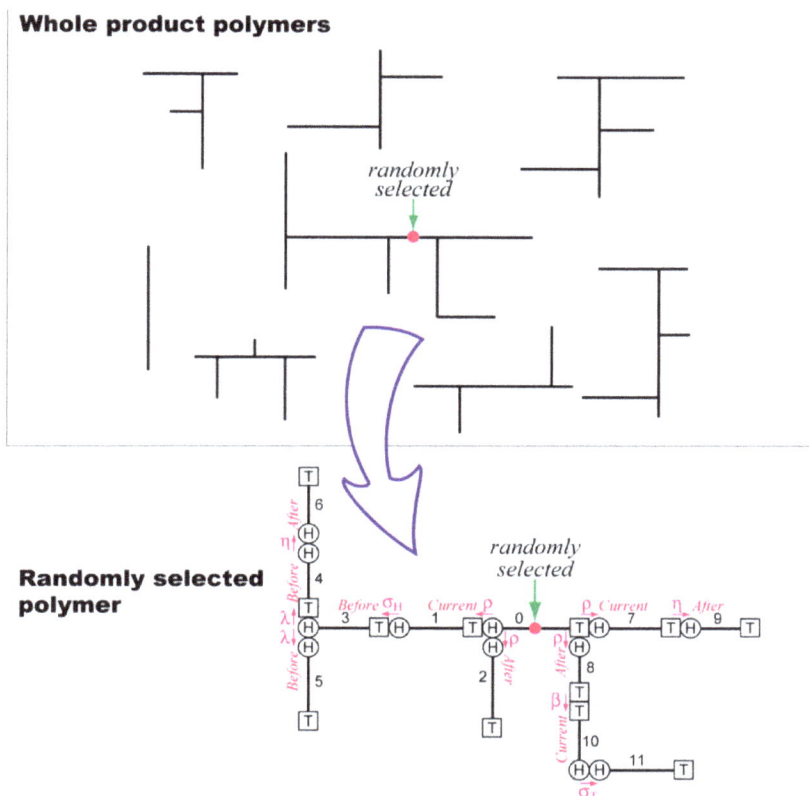

Figure 6. Schematic representation of the concept of random sampling technique.

The selection of polymer molecules can be done both on the number and the weight basis [9]. In the case shown in Figure 6, a polymer molecule is chosen by selecting one monomeric unit randomly, which is the selection on the weight basis. For example, the weight-average degree of polymerization, $\bar{P}_w$ can be obtained by simply taking the arithmetic average of the sample. Incidentally, the analytic representation of $\bar{P}_w$ can always be obtained by using the random sampling technique [20–24].

The elementary reactions considered here are as follows. Initiation (rate represented by R_I), propagation (R_p), chain transfer to small molecules including monomer, solvent and chain transfer agents (R_f), bimolecular termination by disproportionation (R_{td}) and by combination (R_{tc}), and chain transfer to polymer (R_{fp}) leading to long-chain branching (R_b) and chain scission (R_s), with $R_{fp} = R_b + R_s$.

The backbiting reaction to form short-chain branching, typically consisting of several carbon atoms, is not considered explicitly, because it has negligible effects on the formed molecular weight distribution and the radii of gyration, which are the major topics of the present investigation. However, the hydrogen abstraction reaction is much more significant for the tertiary carbon atom, rather than the secondary carbon, and therefore, the polymer transfer reactions would be promoted by the backbiting reactions. Because the location of the tertiary carbon atoms along the chain could be considered random, the overall rate coefficients for the branching and the scission reactions k_b and k_s given below are employed [7,12,18].

$$R_b = k_b[R^\bullet]Q_1 \tag{3}$$

$$R_s = k_s[R^\bullet]Q_1 \tag{4}$$

where $[R^\bullet]$ is the total radical concentration, and Q_1 is the first moment of the polymer distribution, representing the total number of monomeric units in polymer.

$$Q_1 = \sum_{P=1}^{\infty} P\,[P_P] \tag{5}$$

where $[P_P]$ is the concentration of polymer having degree of polymerization, P.

In a strict sense, the long-chain branching reaction shown in Figure 1 is a second order reaction between an internal radical and a monomer, while the scission reaction is the first order reaction of the internal radical, and relative contribution would change during polymerization. However, it was found earlier that such kinetic differences could be neglected [25] for the cases with the final conversion smaller than *ca.* 0.25, which is normally satisfied for the commercial low-density polyethylene production processes.

Actual MC simulation method for the tanks-in-series model is discussed in detail in references [7,18]. In this article, a series of N CSTRs, consisting of one big tank placed as the L-th tank and the same N-1 small tanks as shown in Figure 7 is considered. In the present reaction system, what is important is the magnitude of ξ, defined by the following equation, rather than the volume of reactor [7].

$$\xi_i = k_{p,i}[R^\bullet]_i \bar{t}_i \tag{6}$$

where $k_{p,i}$ and $[R^\bullet]_i$ are the propagation rate constant and the radical concentration both in the ith tank, and $\bar{t}_i$ is the mean residence time of the ith tank, *i.e.*, $\bar{t}_i = V_i/v$. Here, V_i is the volume of the ith tank, and v is the volumetric flow rate. By neglecting the density change, v can be considered as a constant.

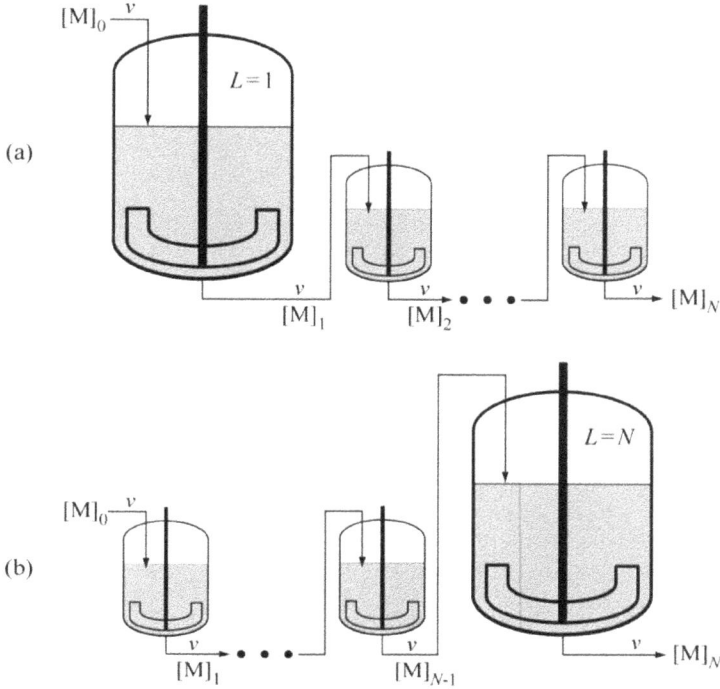

Figure 7. Illustrative representation of the N CSTRs in series, with the arrangement of (**a**) the first large tank $L = 1$; and (**b**) the last large tank $L = N$.

From the mathematical point of view, the cases where one large ξ-value, ξ_L with $\xi_i = \xi$ for $i \neq L$ are discussed in this article. The value of ξ can be controlled by changing $k_{p,i}$ and/or $[R^\bullet]_i$, however by assuming constant values for $k_{p,i}$ and $[R^\bullet]_i$, the control factor is the volume of reactor V, as schematically represented by Figure 7.

Under the condition where only the L-th tank is large with the same volume for the other tanks, because every fluid element must flow through all the tanks, the residence time distribution is kept the same irrespective of the tank arrangement. In the present article, the case where the volume of one large tank is equal to the sum of other tanks, $V_L = (N-1)V$, is considered. In this case, the residence time distribution is shown in Figure 8 [7], where $\theta = t/\bar{t}$, and $\bar{t}$ is the mean residence time of the whole reactor system. The long tail is formed due to a single large CSTR that takes up a half of total reactor volume.

9

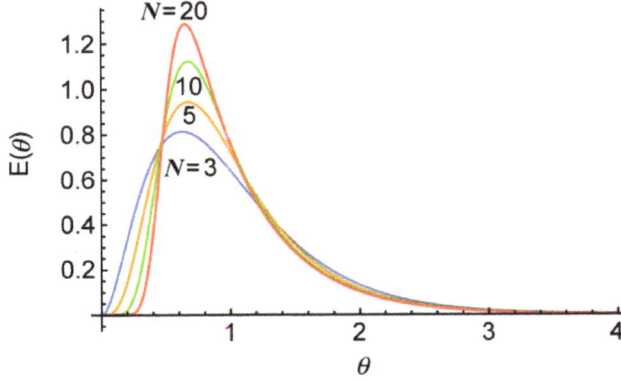

Figure 8. Calculated residence time distribution for the cases with the total number of tanks, N = 3, 5, 10 and 20 [7].

In the present reactor system, the final conversion x_N, the average branching density of the product polymer $\bar{\rho}_N$, and the average scission density of the product polymer $\bar{\eta}_N$ are given, respectively, by [7]:

$$x_N = 1 - \frac{1}{(1 - \xi_L)(1 - \xi)^{N-1}} \tag{7}$$

$$\bar{\rho}_N = \frac{C_b}{x_N (1 + \xi_L)} \left[(\xi_L)^2 + \xi \sum_{i=1}^{N-1} \left\{ (1 + \xi_L) - \left(\frac{1}{1 + \xi} \right)^i \right\} \right] \tag{8}$$

$$\bar{\eta}_N = \frac{C_s}{x_N (1 + \xi_L)} \left[(\xi_L)^2 + \xi \sum_{i=1}^{N-1} \left\{ (1 + \xi_L) - \left(\frac{1}{1 + \xi} \right)^i \right\} \right] \tag{9}$$

where C_b and C_s are defined respectively by $C_b = k_b/k_p$ and $C_s = k_s/k_p$.

Equations (7)–(9) do not involve the large tank number L, which means that x_N, $\bar{\rho}_N$ and $\bar{\eta}_N$ do not change irrespective of the tank arrangement. The present system of CSTRs in series makes it possible to investigate the effect of tank arrangement, while keeping the following properties the same; the final conversion, the final average branching and scission densities, and the residence distribution of the whole reactor system.

3. Results and Discussion

The parameters used in the present article are shown in Table 2. The values of rate ratios, τ and β may be different depending on the tank number, however, they are assumed to be constant in the present investigation. The polymer transfer constant, $C_{fp} = k_{fp}/k_p$ is equal to the sum of C_b and C_s, i.e., $C_{fp} = C_b + C_s$. For C3, combination termination is included. With combination termination, the crosslinks are formed, and possibility of gelation needs to be considered [24]. With C3, the value of C_s is increased to 0.005, in order to prevent gelation.

Table 2. Parameters used in the present investigation.

Parameter	C1	C3
$\tau = (R_{td} + R_f)/R_p$	0.002	0.002
$\beta = R_{tc}/R_p$	0	0.001
$C_b = k_b/k_p$	0.02	0.02
$C_s = k_s/k_p$	0.001	0.005

The MC simulations are conducted to generate 5×10^5 polymer molecules to determine the statistical properties of the product polymers. The total number of tanks, N investigated in this study is $N = 5$ and 10.

Figure 9 shows how the conversion increases with the progress of tank number for $N = 5$. The final conversion is set to be $x_N = 0.2$. For $L = 1$, the conversion goes up in the first tank to $x_1 = 0.1035$ and increases gradually to $x_5 = 0.2$. For $L = 2$, the conversion increases significantly in the second tank. Figure 10 shows the comparison with $N = 10$.

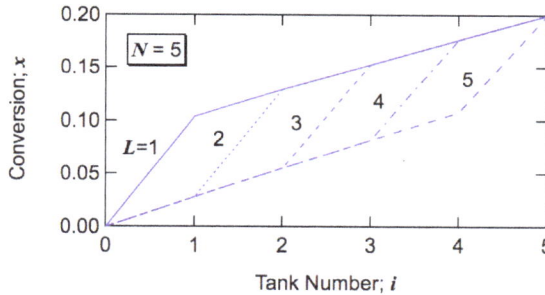

Figure 9. Development of monomer conversion to polymer x with the progress of tank number i for $N = 5$.

Figure 11 shows the development of average branching density $\bar{\rho}$, with the progress of tank number for $N = 5$ and 10. The final average branching density is slightly higher for $N = 5$, although the final conversion is the same. The small difference in $\bar{\rho}_N$ is caused by the series of small tanks part. With $N = 5$, there are only four small tanks, while there are nine small tanks for $N = 10$. Larger number of the same-sized tanks makes the behavior closer to a PFR. When the conversion level is set to be the same, a PFR leads to smaller average branching density than a series of CSTRs, as discussed in Introduction.

11

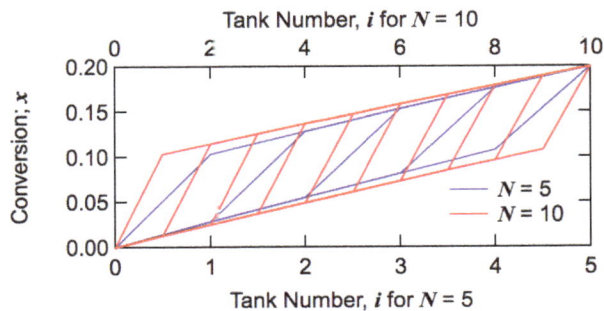

Figure 10. Development of monomer conversion to polymer x with the progress of tank number i for $N = 5$ (blue) and $N = 10$ (red).

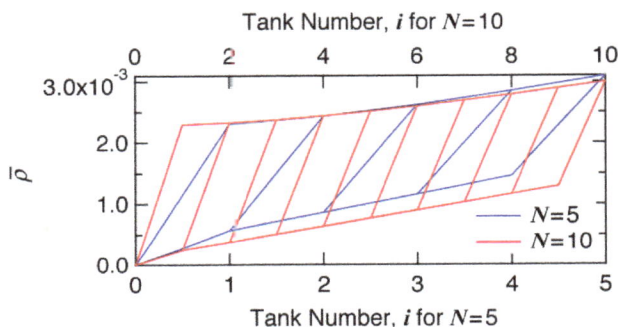

Figure 11. Development of average branching density $\bar{\rho}$ with the progress of tank number i for $N = 5$ (blue) and $N = 10$ (red).

Bearing the developments of conversion and average branching density shown in Figures 10 and 11 in mind, let us investigate the effect of tank arrangement on the formed branched structure.

3.1. Molecular Weight Distribution

Figure 12 shows the final average degree of polymerization formed in each reactor system. The symbols are MC simulation results. The red lines and symbols are for $N = 5$, and the blue ones are for $N = 10$. The weight-average degree of polymerization, $\overline{P}_w$ is the largest for the first large tank case ($L = 1$), and decreases with larger L-values.

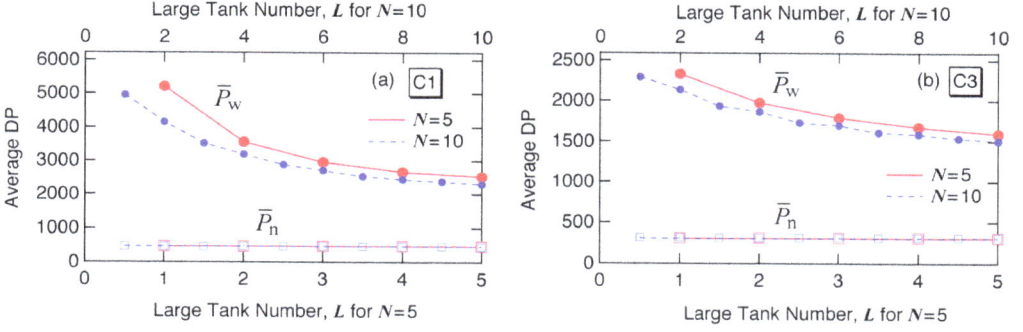

Figure 12. MC simulation results for the number ($\overline{P}_n$) and weight-average ($\overline{P}_w$) degree of polymerization of the polymers formed for conditions (**a**) C1 and (**b**) C3.

The number-average degree of polymerization, $\overline{P}_n$ is given theoretically by [18]:

$$\overline{P}_n = \frac{1}{\tau + \beta/2 + \overline{\eta}_N} \tag{10}$$

Because the final average scission density $\overline{\eta}$, as well as the values of τ and β, is the same irrespective of tank arrangement, $\overline{P}_n$ does not change with the large tank number L, as Equation (10) and the MC simulation results show.

Figure 13 shows the full weight fraction distribution for $N = 10$. The cases with $N = 5$ can be found in ref. [7]. The high molecular weight tail prevails in $L = 1$.

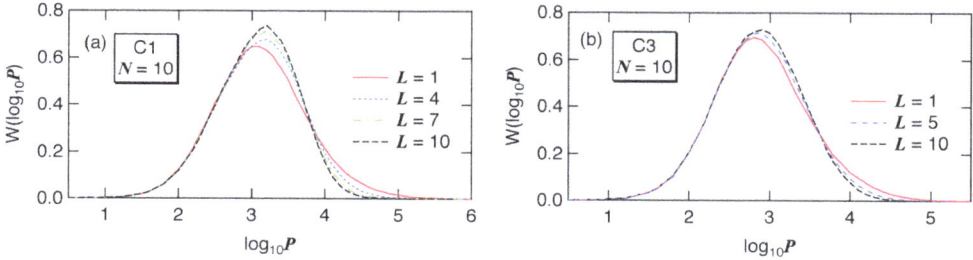

Figure 13. MC simulation results for the weight fraction distribution, plotted as a function of $\log_{10}P$, $W(\log_{10}P)$ of the polymers formed for conditions (**a**) C1 and (**b**) C3, with $N = 10$.

Now, consider the qualitative explanation why the first big tank case gives the largest $\overline{P}_w$. Two important characteristics need to be reminded. The long-chain branches are formed through the polymer transfer reaction, and larger polymer molecules have a better chance of being attacked by a polymer radical. Therefore, (1) larger molecules grow faster than the smaller molecules in the present reaction system. Note that the frequency of long-chain branching is higher than that of chain scission, $C_b > C_s$ in the present reaction system. (2) A CSTR produces broader molecular weight distribution (MWD) than that for a PFR, as shown in Figure 2. In a CSTR, the polymer molecules whose residence time is large tend to grow significantly.

Now, consider the case with $L = 1$. In this case, large polymer molecules are formed in the first big CSTR. After that all the polymers flow through a series of small CSTRs, which resembles a PFR. According to Item (1), larger polymer molecules grow faster in a PFR, and very large polymer molecules can be formed for the case with $L = 1$.

On the other hand, for the case with the last big tank, $L = N$, the first part is a PFR. With a PFR, the MWD surely becomes broader during polymerization, but not very much compared with a CSTR. In the final big CSTR, the MWD becomes broader, however, large polymer molecules formed before entering the last big CSTR do not necessarily stay long time in this CSTR and may not grow significantly. Therefore, the last CSTR is not very effective to make larger molecules even larger. As a result, $\overline{P}_w$ is small for the case with the last big tank, $L = N$.

3.2. Branching Density

Figure 14 shows the MC simulation results for the relationship between the branching density ρ and the degree of polymerization P for $N = 10$. Figure 14a–c are for condition C1, and Figure 14d–f are for condition C3. The MC simulation results for $N = 5$ can be found in ref. [7], and the fundamental characteristics are the same as the present simulation results for $N = 10$. In Figure 14, each dot represents ρ and P of each polymer molecule simulated, and the solid blue curve with circular symbols shows the average branching density of polymers having degree of polymerization P, $\overline{\rho}(P)$. The dashed black line shows the average branching density of the whole system $\overline{\rho}$, which is $\overline{\rho} = 0.003$ for all cases shown in Figure 14. As in the cases of other branched polymer systems, the value of $\overline{\rho}(P)$ increases with P but reaches a constant limiting value $\rho_{P \to \infty}$, and $\rho_{P \to \infty}$ is larger than the average branching density of the whole system, $\overline{\rho}$. Because the $\overline{\rho}$-value, shown by the broken line is the same, one notices that $\rho_{P \to \infty}$ becomes larger as the value of L increases, i.e., as the big tank moves backward.

As shown in Figure 12, the weight-average DP, $\overline{P}_w$ becomes smaller as the value of L increases. According to the present series of simulation results, $\rho_{P \to \infty}$ is larger for the cases that gives smaller $\overline{P}_w$. This is a completely opposite result, compared with the relationship between a PFR and a CSTR, as shown in Figures 2 and 3. A CSTR gives larger $\rho_{P \to \infty}$ and larger $\overline{P}_w$, compared with a PFR. Now, think of the qualitative explanation for this seemingly strange conflict.

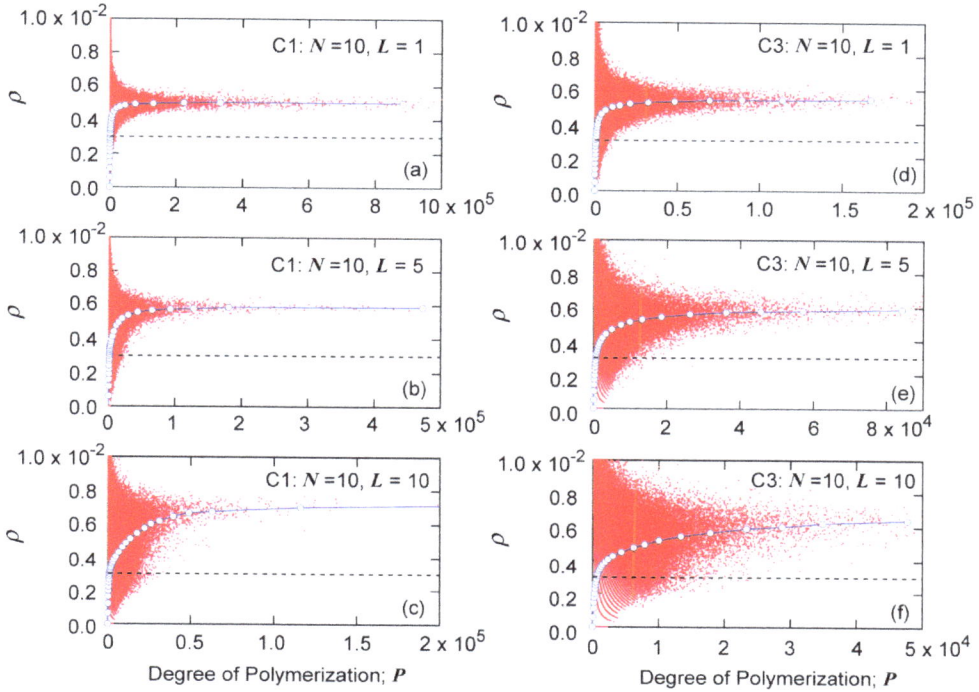

Figure 14. Relationship between branching density ρ and degree of polymerization P, obtained from the MC simulation for $N = 10$: (**a**) C1, $L = 1$; (**b**) C1, $L = 5$; (**c**) C1, $L = 10$; (**d**) C3, $L = 1$; (**e**) C3, $L = 5$; and (**f**) C3, $L = 10$. The black broken line shows the average branching density of the whole reaction system, $\bar{\rho} = 0.003$ for all cases.

First, consider the $\bar{\rho}(P)$ curves of a CSTR and a PFR shown in Figure 3c. A larger branching density for a given degree of polymerization (P) simply means that the average chain length of the primary chains that make up a branch polymer molecule having P is smaller for a CSTR. In free-radical polymerization, each primary chain is formed within a very small time interval, and therefore, one can conveniently define the instantaneous chain length distribution [8,9]. The number-average chain length of a linear polymer radical formed through propagation starting from the time of radical generation until the dead chain formation is given by [9]:

$$\overline{P}^{\bullet}_{\text{n,linear}} = \frac{R_{\text{p}}}{R_{\text{td}} + R_{\text{f}} + R_{\text{tc}} + R_{\text{fp}}} = \frac{1}{\tau + \beta + R_{\text{fp}}/R_{\text{p}}} \tag{11}$$

15

In a PFR, the final term, R_{fp}/R_p is 0 at $x = 0$ because the polymer concentration is 0, and increases to the value at the final conversion, x_f. For a CSTR, R_{fp}/R_p is a constant, and is given by:

$$\frac{R_{fp}}{R_p} = C_{fp}\frac{x_f}{1 - x_f} \tag{12}$$

where C_{fp} is the polymer transfer constant, and is equal to $C_{fp} = C_b + C_s$. Equation (12) represents the maximum value for a PFR reached at x_f.

The primary chain length is smaller for a CSTR. A larger number of branch points are required to form a branched polymer molecule having a given degree of polymerization, P. This is the reason for the larger $\bar{\rho}(P)$ curve of a CSTR, compared with a PFR, shown in Figure 3c.

Next, consider the CSTRs in series, with $L = 1$ and $L = N$. As shown in Figure 10, about one half of polymer by weight is formed in the large tank, and therefore, the L-th CSTR has the most important effect on the formed branched structure. The value of R_{fp}/R_p is $C_{fp}x_1/(1-x_1)$ for $L = 1$, and is $C_{fp}x_N/(1-x_N)$ for $L = N$. Because $x_N > x_1$, R_{fp}/R_p is larger for $L = N$, leading to give smaller primary chains for $L = N$. Therefore, a larger number of branch points are required to form branched polymers having a given P. The $\rho_{P\to\infty}$ is larger for $L = N$ that gives smaller $\bar{P}_w$.

3.3. Radius of Gyration

Figure 15 shows the MC simulation results for the relationship between the mean-square radius of gyration $<s^2>_0$ and degree of polymerization P of the polymer molecules formed in various reactor placements with $N = 10$. Similar figures for $N = 5$ can be found in ref. [7]. Each dot in Figure 15 represents $<s^2>_0$ and P-value of each polymer molecule simulated, and the blue curve with circular symbols shows the average $<s^2>_0$ of polymers having DP, P. The black curve shows the radius of gyration of the random branched polymers given by the Zimm-Stockmayer equation [16], *i.e.*, Equation (1). The radius of gyration is smaller than that for random branched polymers, but the degree of deviation becomes smaller as the value of L increases.

As was shown in Figure 4b, much more compact polymers are formed in a CSTR. In a CSTR, the primary polymer molecules whose residence time is large are expected to possess a larger number of branch points. These primary chains tend to form a core region of a star-like structure. This is the reason for forming compact branched polymers, especially for large polymers. On the other hand, Figure 4a shows that more randomly branched structure is tend to be formed in a PFR.

In the CSTRs in series with $L = 1$, compact star-like polymers are formed in the first big CSTR. After the first tank, branches are attached to the star-like polymers rather randomly, and the star-like structure is preserved, leading to form polymers with smaller $<s^2>_0$, compared with that for random branched polymers.

16

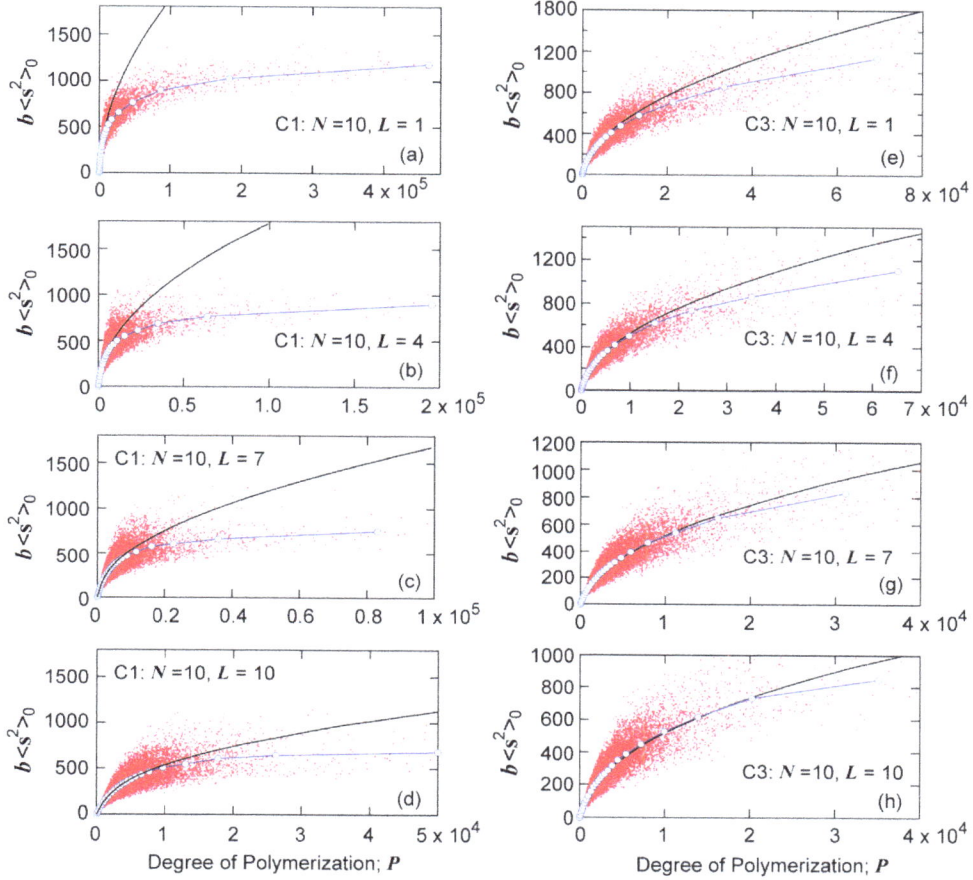

Figure 15. Relationship between the mean-square radius of gyration $<s^2>_0$ and degree of polymerization P of the polymer molecules formed in various reactor placements with $N = 10$: (**a**) C1, $L = 1$; (**b**) C1, $L = 4$; (**c**) C1, $L = 7$; (**d**) C1, $L = 10$; (**e**) C3, $L = 1$; (**f**) C3, $L = 4$; (**g**) C3, $L = 7$; and (**h**) C3, $L = 10$. The black curve shows the relationship for random branched polymers calculated from Equation (1).

Now, consider the case with the last big tank, $L = N$. In the first stage of N-1 small tanks, the reactor characteristics are more like a PFR, and polymers with relatively random branched structure are expected to be formed. In the final big tank, the branched polymer molecules whose residence time is large will connect many branch chains. However, looking from an entering branched polymer molecule, the branch points formed on the polymer molecule would be distributed rather randomly, and a core region will not be formed. This is the reason for obtaining polymers having relatively randomly branched polymers.

With $L = 1$, the branching density, $\rho_{P\to\infty}$ is small, as shown in Figure 14a, but much more compact polymers are formed, compared with the random branched polymers, as shown in Figure 15a. On the other hand, with $L = N$, the branching density, $\rho_{P\to\infty}$ is large, as shown in Figure 14c, but the branched structure is more random and $<s^2>_0$ is large for the given branching density level, as shown in Figure 15d. As a result, the $<s^2>_0$-value for the given P becomes essentially the same, irrespective of the tank arrangement, as shown in Figure 16. Similarly as reported in reference [7], there seem to exist small differences for C1, but the differences are very small for C3.

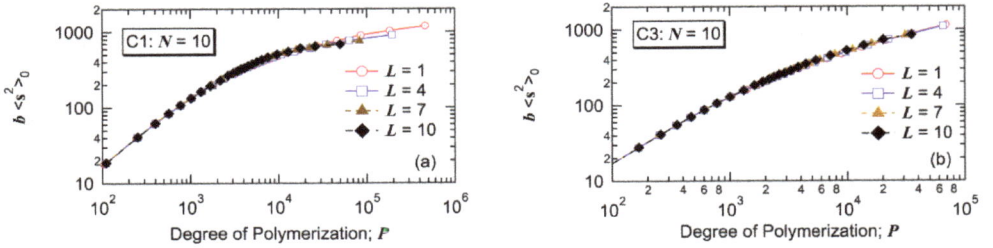

Figure 16. Relationship between the mean-square radius of gyration $<s^2>_0$ and degree of polymerization P of the polymer molecules formed with $N = 10$: (**a**) C1 and (**b**) C3.

4. Conclusions

In this article, a series of continuous stirred-tank reactors, consisting of one big tank and the same $N-1$ small tanks is investigated to design a new reactor system for free-radical polymerization with simultaneous long-chain branching and minor contribution of scission. Table 3 summarizes the results obtained in this study. When the final conversion is set to be the same, the present reactor configuration does not change the final average branching and scission densities, irrespective of the tank arrangement. The residence time distribution does not change with the tank arrangement, and therefore, the differences are not caused by the residence time distribution, but by the different history of branched polymer formation.

The weight-average DP, $\overline{P}_w$ is the largest for the first big tank case $L = 1$, and decreases as the big tank moves backward, *i.e.*, as L increases. The branching density of large polymers, $\rho_{P\to\infty}$ is the smallest for $L = 1$, and becomes larger as L increases. With $L = 1$, polymers with star-like globular structure are formed. By increasing L, the polymer structure changes to a more randomly branched structure.

With $L = 1$, the branching density of large polymers ($\rho_{P\to\infty}$) is small, while the structure is star-like and $<s^2>_0$-value is small for the given branching density. On the other hand, with $L = N$, the branching density of large polymers ($\rho_{P\to\infty}$) is large, while the structure is more like randomly branched and $<s^2>_0$-value is large for the given branching density level. What is interesting is that the relationship between $<s^2>_0$ and P is essentially unchanged by the tank arrangement.

Table 3. N CSTRs in series with one large tank, when the final conversion, x_N, is set to be the same.

Large Tank Number, $L = 1 \ldots N$	
Final average branching density	Unchanged
Final average scission density	Unchanged
Residence time distribution	Unchanged
Weight-average degree of polymerization, $\overline{P}_w$	>
Branching density for large polymers, $\rho_{P \to \infty}$	<
Branched structure	Star-like $\leftrightarrow$ Random
Relationship between $<s^2>_0$ and P	Essentially the same

The present investigation shows that by changing the tank arrangement, various types of branched polymers, from star-like globular structure to a more randomly branched structure, can be obtained, while keeping the followings are the same: (1) final conversion; (2) final average branching and scission densities; and (3) the relationship between $<s^2>_0$ and P. In the case of a tower-type multi-zone reactor, the branched structure could be controlled by changing the location of division plates. On the basis of the present investigation, it is expected that the first big zone case will tend to produce polymers with star-like globular structure, while the case with the last big zone arrangement will tend to form more randomly branched polymers.

Conventionally, most mathematical models have been used to describe the behavior in the reactor that has been applied already in industry. The simulation has been used rather passively. The present type of positive use of simulation may lead to propose novel reactor systems. Table 4 highlights the characteristics of the positive model use adopted in this article.

Table 4. Positive model use proposed in this article.

Passive Model Use	Positive Model Use
Aiming at improving the process already used in industry	Aiming at creating a new process
Emphasizing the quantitative agreement with the experimental data	Emphasizing the novelty of the proposed processes
Straightforward cause-and-effect logic	Divergence through try-and-error and convergence through reasoning
Initiative by chemists	Initiative by process engineers

Acknowledgments: I would like to express my sincere thanks to Professor Masoud Soroush, Drexel University, for kind invitation to this special issue.

Conflicts of Interest: The author declares no conflict of interest.

References

1. Ehrlich, P.; Mortimer, G.A. Fundamentals of the free-radical polymerization of ethylene. *Adv. Polym. Sci.* **1970**, *7*, 386–448.
2. Goto, S.; Yamamoto, K.; Furui, S.; Sugimoto, M. Computer model for commercial high-pressure polyethylene reactor based on elementary reaction rates obtained experimentally. *J. Appl. Polym. Sci. Appl. Polym. Symp.* **1981**, *36*, 21–40.
3. Giudici, R.; Hamielec, A.E. A simulation study on random scission of branched chains. *Polym. React. Eng.* **1996**, *4*, 73–101.
4. Tobita, H. Random degradation of branched polymers. 2. multiple branches. *Macromolecules* **1996**, *29*, 3010–3021.
5. Iedema, P.D.; Wulkow, M.; Hoefsloot, C.J. Modeling molecular weight and degree of branching distribution of low-density polyethylene. *Macromolecules* **2000**, *33*, 7173–7184.
6. Iedema, P.D. Predicting MWD and branching distribution of terminally branched polymers undergoing random scission. *Macromol. Theory Simul.* **2012**, *21*, 166–186.
7. Tobita, H. Continuous tanks-in-series process for free-radical polymerization with long-chain branching and scission: Effect of the order of a large tank. *Macromol. React. Eng.* **2015**.
8. Tobita, H. Random sampling technique to predict the molecular weight distribution in nonlinear polymerization. *Macromol. Theory Simul.* **1996**, *5*, 1167–1194.
9. Tobita, H. Polymerization processes, 1. Fundamentals. In *Ullmann's Encyclopedia of Industrial Chemistry*; Wiley-VCH: Weinheim, Germany, 2015.
10. Tobita, H. Simultaneous long-chain branching and random scission: 1. Monte Carlo simulation. *J. Polym. Sci. Part B: Polym. Phys.* **2001**, *39*, 391–403.
11. Tobita, H.; Kawai, H. Simulation of size exclusion chromatography for branched polymers formed by simultaneous long-chain branching and random scission. *E-Polymers* **2002**, *2*, 688–700.
12. Tobita, H. Free-radical polymerization with long-chain branching and scission in a continuous stirred-tank reactor. *Macromol. React. Eng.* **2013**, *7*, 181–192.
13. Read, D.J.; Auhl, D.; Das, C.; den Doelder, J.; Kapnistos, M.; Vittorias, I.; McLeish, T.C.B. Linking models of polymerization and dynamics to predict branched polymer structure and flow. *Science* **2011**, *333*, 1871–1874. PubMed]
14. Levenspiel, O. *Chemical Reaction Engineering*; John Wiley & Sons: New York, NY, USA, 1972.
15. Tobita, H. Markovian approach to self-condensing vinyl polymerization: Distributions of molecular weights, degrees of branching, and molecular dimensions. *Macromol. Theory Simul.* **2015**, *24*, 117–132.
16. Zimm, B.H.; Stockmayer, W.H. The dimensions of chain molecules containing branches and rings. *J. Chem. Phys.* **1949**, *17*, 1301–1314.

17. Tobita, H. Dimensions of branched polymers formed in simultaneous long-chain branching and random scission. *J. Polym. Sci.: Part B: Polym. Phys.* **2001**, *39*, 2960–2968.
18. Tobita, H. Continuous free-radical polymerization with long-chain branching and scission in a tanks-in-series model. *Macromol. Theory Simul.* **2014**, *23*, 182–197.
19. Brandao, A.L.; Soares, J.B.P.; Pinto, J.C.; Alberton, A.L. When Polymer reaction engineers play dice: Application of monte carlo models in PRE. *Macrtomol. React. Eng.* **2015**, *9*, 141–185.
20. Tobita, H. Molecular weight distribution in random branching of polymer chains. *Macromol. Theory Simul.* **1996**, *5*, 129–144.
21. Tobita, H.; Zhu, S. Statistical crosslinking of heterochains. *Polymer* **1997**, *38*, 5431–5439.
22. Tobita, H. General matrix formula for the weight-average molecular weights of crosslinked polymer systems. *J. Polym. Sci.: Part B: Polym. Phys.* **1998**, *36*, 2423–2433.
23. Tobita, H. Markovian approach to nonlinear polymer formation: Free-radical crosslinking copolymerization. *Macromol. Theory Simul.* **1998**, *7*, 675–684.
24. Tobita, H. Free-radical polymerization with long-chain branching and scission: Markovian solution of the weigh-average molecular weight. *Macromol. Theory Simul.* **2014**, *23*, 477–489.
25. Tobita, H. Markovian approach to free-radical polymerization with simultaneous long-chain branching and scission: Effect of branching and scission kinetics. *Macromol. React. Eng.* **2015**, *9*, 245–258.

Optimal Design for Reactivity Ratio Estimation: A Comparison of Techniques for AMPS/Acrylamide and AMPS/Acrylic Acid Copolymerizations

Alison J. Scott, Marzieh Riahinezhad and Alexander Penlidis

Abstract: Water-soluble polymers of acrylamide (AAm) and acrylic acid (AAc) have significant potential in enhanced oil recovery, as well as in other specialty applications. To improve the shear strength of the polymer, a third comonomer, 2-acrylamido-2-methylpropane sulfonic acid (AMPS), can be added to the pre-polymerization mixture. Copolymerization kinetics of AAm/AAc are well studied, but little is known about the other comonomer pairs (AMPS/AAm and AMPS/AAc). Hence, reactivity ratios for AMPS/AAm and AMPS/AAc copolymerization must be established first. A key aspect in the estimation of reliable reactivity ratios is design of experiments, which minimizes the number of experiments and provides increased information content (resulting in more precise parameter estimates). However, design of experiments is hardly ever used during copolymerization parameter estimation schemes. In the current work, copolymerization experiments for both AMPS/AAm and AMPS/AAc are designed using two optimal techniques (Tidwell-Mortimer and the error-in-variables-model (EVM)). From these optimally designed experiments, accurate reactivity ratio estimates are determined for AMPS/AAm ($r_{AMPS} = 0.18$, $r_{AAm} = 0.85$) and AMPS/AAc ($r_{AMPS} = 0.19$, $r_{AAc} = 0.86$).

Reprinted from *Processes*. Cite as: Scott, A.J.; Riahinezhad, M.; Penlidis, A. Optimal Design for Reactivity Ratio Estimation: A Comparison of Techniques for AMPS/Acrylamide and AMPS/Acrylic Acid Copolymerizations. *Processes* **2015**, *3*, 749–768.

1. Introduction

Some of the most common acrylamide-based copolymer systems used in enhanced oil recovery (EOR) are acrylamide (AAm) and acrylic acid (AAc) copolymers. However, these AAm/AAc copolymers, like many other water-soluble polymers with high molecular weights, are very shear sensitive. That is, when the copolymer is subjected to high temperatures and stresses, there is potential for the polymer backbone to break [1]. This directly affects the polymer's efficiency in enhanced oil recovery, as the polymer in this case will not be able to increase the aqueous phase viscosity as much as was originally desired. Thus, it is essential to minimize polymer degradation in EOR applications.

2-acrylamido-2-methylpropane sulfonic acid (AMPS) has the potential to improve main chain stability in harsh environments. The steric hindrance provided by the sulfonic group in AMPS is expected to control potential degradation of the polymer backbone [2], enhance thermal stability [3], and improve the polymer's resistance to precipitation by limiting hydrolysis [4]. A survey of existing (yet unreliable) reactivity ratios in the literature for the related copolymers (AMPS/AAm and AMPS/AAc) confirms that synthesis and testing of the AMPS/AAm/AAc terpolymer is promising. To tailor-make a water-soluble terpolymer of AMPS/AAm/AAc, polymerization kinetics for the binary components must first be understood. AAm/AAc copolymerization kinetics have recently been clarified [5], so the current study focuses on AMPS/AAm and AMPS/AAc.

The statistically correct error-in-variables-model (EVM) is used for analysis, as it is a non-linear estimation technique that considers the error present in all variables [6,7]. Through EVM and direct numerical model integration, we are also able to estimate reactivity ratios using cumulative composition data (as opposed to standard analysis of low-conversion data). This provides additional advantages, including eliminating unnecessary assumptions and avoiding the experimental challenges associated with collecting low-conversion data [8]. Copolymerizations of both AMPS/AAm and AMPS/AAc are designed using Tidwell-Mortimer (T-M) and error-in-variables-model (EVM) techniques. Reactivity ratios (and associated joint confidence regions) obtained through the traditional T-M design are contrasted with those obtained through EVM design. This allows for a direct comparison between the T-M and EVM design approaches (not readily available in the literature).

1.1. Copolymerization Kinetics

The Mayo-Lewis model is widely used for copolymerization systems. This classical equation, also called the instantaneous copolymer composition equation, is presented in Equation (1).

$$\frac{d\,[M_1]}{d\,[M_2]} = \left(\frac{[M_1]}{[M_2]}\right)\left(\frac{r_1\,[M_1] + [M_2]}{[M_1] + r_2\,[M_2]}\right) \tag{1}$$

where $[M_1]$ and $[M_2]$ are the concentrations of monomer 1 and 2 in the polymerizing mixture, and

$$r_1 = \frac{k_{p11}}{k_{p12}} \text{ and } r_2 = \frac{k_{p22}}{k_{p21}} \tag{2}$$

The monomer reactivity ratios, r_1 and r_2, describe the potential for homo-propagation relative to cross-propagation. These parameters are specific to each copolymer system, and many summary tables are available citing reactivity ratios of common copolymer systems [9]. Reactivity ratios can be estimated using experimental data, if the free (unreacted) monomer composition in the polymerizing mixture and the bound (incorporated) monomer composition in the polymer chains (*i.e.*, copolymer composition) are known.

Another popular form of the copolymerization equation (Equation (1)) is given by Equation (3), which provides information directly about the instantaneous composition of the copolymer, F_1, given the comonomer composition in the polymerizing mixture.

$$F_1 = \frac{r_1 f_1^2 + f_1 f_2}{r_1 f_1^2 + 2f_1 f_2 + r_2 f_2^2} \tag{3}$$

where f_1 and f_2 represent the mole fractions of unreacted monomer 1 and monomer 2 in the polymerizing mixture. F_1 is the instantaneous mole fraction of monomer 1 units bound (incorporated) in the copolymer chains, corresponding to f_1.

An additional point of interest in copolymerization kinetics is establishing the azeotropic composition (if it exists) for the system. At the azeotropic point, the feed composition (f_1) and the instantaneous copolymer composition (F_1) are equivalent. If the reactivity ratios are known, we can use the instantaneous copolymerization equation (Equation (3)) to examine F_1 as a function of f_1 and to establish the azeotropic point. By setting $F_1 = f_1$, Equation (3) is simplified to the binary azeotropic composition, shown in Equation (4) [10].

$$F_1 = f_1 = \frac{1 - r_2}{2 - r_1 - r_2} \tag{4}$$

Determination of azeotropic composition is just one application for reactivity ratios, which are extremely important parameters for copolymerization kinetics. Reactivity ratios can also be used to predict polymer properties such as copolymer composition or sequence length, and could eventually be used in custom polymer production for specific applications [11]. Therefore, it is essential that reactivity ratio estimates be as accurate as possible. Techniques for reactivity ratio estimation are briefly discussed in what follows.

1.2. Reactivity Ratio Estimation

In general, reactivity ratios are parameters obtained from experimental data by analyzing the copolymer composition at several different feed compositions. Traditionally, linear regression techniques have incorrectly been used for reactivity ratio estimation. These techniques include the Mayo-Lewis method (method of intersections), the Fineman-Ross method and the Kelen-Tudos method [12]. These techniques were originally chosen for their simplicity, as technology was not readily available for intense computation. However, linearizing the kinetic models (which are inherently non-linear in the parameters) requires making imprecise, subjective and invalid assumptions. An additional consideration is the use of the instantaneous copolymerization model in these linear techniques; the reaction must be kept at low conversion so that the assumption of "constant composition" in the feed is somewhat valid [8]. However, polymerizations at low conversions are extremely error-prone, and it is impossible to guarantee that the feed composition will remain constant (especially when dealing with an unstudied system).

1.3. Design of Experiments

Optimal design of experiments leads to increased information content while minimizing the number of experiments and obtaining more precise parameter estimates [7]. Tidwell and Mortimer [13] applied an (approximate) D-optimality criterion to the Mayo-Lewis copolymerization equation to determine the best monomer feed compositions at which to run reactivity ratio estimation experiments:

$$f_{2,1} = \frac{r_1}{2 + r_1} \quad \text{and} \quad f_{2,2} = \frac{2}{2 + r_2} \tag{5}$$

where $f_{2,1}$ and $f_{2,2}$ denote the initial feed composition of monomer 2 for the first and second experiments, respectively. Preliminary reactivity ratio estimates (r_1 and r_2) can be obtained from the literature or from some type of preliminary experimentation.

D-optimality is an extremely powerful criterion, and through its "ease-of-use" it can act as a good starting point for experimental design. A more complex, yet equally valid, technique for designing optimal reactivity ratio estimation experiments is the error-in-variables-model (EVM) [14]. EVM is not only used for reactivity ratio estimation, but also employed in the preceding design of experiments stage; the technique considers error terms in all variables involved (both independent and dependent) in the process model.

2. Experimental

2.1. Reagent Purification

Monomers 2-acrylamido-2-methylpropane sulfonic acid (AMPS; 99%), acrylamide (AAm; electrophoresis grade, 99%), and acrylic acid (AAc; 99%) were purchased from Sigma-Aldrich (Oakville, ON, Canada). AAc was purified via vacuum distillation at 30 °C, while AAm and AMPS were used as received. Initiator (4,4'-azo-bis-(4-cyanovaleric acid), ACVA), inhibitor (hydroquinone) and sodium hydroxide were also purchased from Sigma-Aldrich. Sodium chloride from EMD Millipore (Etobicoke, ON, Canada) was used as received. In terms of solvents, water was Millipore quality (18 MΩ·cm); acetone (99%) and methanol (99.8%) were used as received from suppliers. Nitrogen gas (4.8 grade) used for degassing solutions was purchased from Praxair (Mississauga, ON, Canada).

2.2. Polymer Synthesis

In general, the experimental techniques described by Riahinezhad *et al.* [5] were adopted for these copolymer systems. Monomer solutions with a total monomer concentration of 1 M were prepared. The comonomer ratios in each system (AMPS/AAm and AMPS/AAc) are described in detail later as part of the experimental design for each individual system. The monomer solutions were titrated with sodium hydroxide to adjust the pH to approximately 7 ($\pm$0.5). Each recipe had 0.004 M initiator (ACVA), and sodium chloride was added to ensure constant ionic strength among the experiments. Constant pH and ionic strength are extremely important in copolymer and terpolymer synthesis, as has been demonstrated previously [15]. The solutions were then purged with 200 mL/min nitrogen for 2 h. After degassing, aliquots of ~20 mL of solution were transferred to sealed vials using the cannula transfer method [5]. Free-radical solution (aqueous phase) polymerizations were run in a temperature controlled shaker-bath (OLS200; Grant Instruments, Cambridge, UK) at 40 °C and 100 rpm. Vials were removed at selected time intervals, placed in ice and further injected with approximately 1 mL of 0.2 M hydroquinone solution to stop the polymerization. Polymer samples were isolated by precipitating the products in acetone or methanol, filtered (paper filter grade number 41, Whatman; Sigma-Aldrich, Oakville, ON, Canada) and vacuum dried for 1 week at 50 °C. All polymerizations were independently replicated.

2.3. Polymer Characterization

Conversion of the polymer samples was determined using gravimetry. The mass of the sodium ions was also considered in conversion calculations, as per the recommendation of Riahinezhad *et al.* [15]. Copolymer composition was measured using elemental analysis (CHNS, Vario Micro Cube, Elementar). Calculation of composition did not include H content, as residual water has been known to affect the determined H content [5]. Select samples were independently replicated.

3. AMPS/AAm Copolymer

3.1. Literature Background for AMPS/AAm

The majority of the work in the copolymerization of 2-acrylamido-2-methylpropane sulfonic acid (AMPS) with acrylamide (AAm) has focused on crosslinking systems, as crosslinked copolymers of AMPS and AAm have applications as superabsorbent hydrogels (e.g., see [16–19]). As with many other copolymer systems, such studies look at the final polymer (synthesis and characterization without

26

considering the full conversion trajectory) and its performance properties, while they rarely investigate polymerization kinetics or reactivity ratio estimation. There has also been some work done in examining the effectiveness of AMPS/AAm copolymers in enhanced oil recovery (EOR) [2,20–22]. The focus of these articles is intended to be the synthesis and testing of polymers for EOR use.

The objective here is to obtain accurate and reliable reactivity ratios for the AMPS/AAm copolymer. Therefore, Table 1 provides a summary of reactivity ratios as reported in the literature for the copolymerization of AMPS and AAm. Although some of the estimates are similar (especially for r_{AAm}), there are evident inconsistencies between experimental techniques and reactivity ratio estimation methods. It is also important to note that all of the estimation techniques used to date have been linear. Given the numerous sources of error associated with linear estimation methods and the advantages of non-linear techniques, it seems only reasonable that future reactivity ratios be estimated using EVM [7].

Table 1. Reactivity ratio summary for AMPS/AAm.

Ref.	Experimental	Estimation Technique	r_{AMPS}	r_{AAm}
[16]	–Type: Aqueous solution crosslinking copolymerization –Initiator: KPS –Temperature: 40 °C –pH = 7 –Composition: IR and EA	Comparison of feed and copolymer compositions (no statistical estimation)	1.00	1.00
[23]	–Type: Aqueous solution copolymerization –Initiator: KPS –Temperature: 50 °C –Composition: EA	Billmeyer * [24] Billmeyer * [24] Kelen-Tudos Average	0.76 0.70 0.62 0.70 ± 0.08	1.00 1.06 1.21 1.10 ± 0.10
[25]	–Type: Aqueous solution copolymerization –Initiator: KPS –Temperature: 35 °C and 55 °C –Composition: H-NMR and vibrational Raman spectroscopy	Fineman-Ross	1.00	1.00
[26]	–Type: Aqueous solution copolymerization –Initiator: KPS –Temperature: 30 °C –pH = 9 –Composition: IR and EA	Fineman-Ross Kelen-Tudos Integrated Mayo-Lewis	0.49 ± 0.02 0.52 ± 0.07 0.50 ± 0.01	0.98 ± 0.09 1.00 ± 0.08 1.02 ± 0.01
[27]	–Type: Aqueous solution copolymerization –Initiator: APS –Temperature: 60 °C –Composition: EA and C-NMR	Fineman-Ross Kelen-Tudos	0.37 ± 0.04 0.42 ± 0.03	1.01 ± 0.01 1.05 ± 0.06
[27]	–Type: Aqueous solution redox copolymerization –Initiator: APS/NaHSO₃ –Temperature: 25 °C –Composition: C-NMR	Fineman-Ross Kelen-Tudos	0.54 ± 0.03 0.51 ± 0.03	1.07 ± 0.01 1.05 ± 0.06

Nomenclature: AAm, acrylamide; AMPS, 2-acrylamido-2-methylpropane sulfonic acid; APS, ammonium persulfate; EA, elemental analysis; IR, infrared spectroscopy; KPS, potassium persulfate; NMR, nuclear magnetic resonance; * Note: Based on estimation approaches described in Billmeyer [24].

27

3.2. Design of Experiments for AMPS/AAm

Both the Tidwell-Mortimer (T-M) and error-in-variables-model (EVM) design of experiments rely on preliminary reactivity ratio estimates. Therefore, experimental work begins with preliminary experiments, which are based on existing literature values from McCormick and Chen ($r_{AMPS} = 0.50$, $r_{AAm} = 1.02$) [26]. Once preliminary reactivity ratio estimates are established, the T-M and EVM criteria can be used to design optimal experiments. Each experimental design provides two feed compositions (in terms of monomer 1; AMPS in this case) at which to run new experimental trials, and the results are presented below. In Table 2, $f_{AMPS,0,1}$ represents the first initial feed composition (in terms of AMPS) from the design, just as $f_{AMPS,0,2}$ represents the second initial feed composition from the design. The reactivity ratio estimates obtained from each design are also included for easy comparison. More details on the determination of reactivity ratio estimates follow.

Table 2. Design of experiments and reactivity ratio estimates for AMPS/AAm.

Approach	Reactivity Ratios for Design		Feed Compositions (Mole Fractions)		New Reactivity Ratio Estimates	
	r_{AMPS}	r_{AAm}	$f_{AMPS,0,1}$	$f_{AMPS,0,2}$	r_{AMPS}	r_{AAm}
Preliminary	0.50	1.02	0.15	0.80	0.13	0.84
T-M Design	0.13	0.84	0.30	0.91	0.16	0.77
EVM Design	0.13	0.84	0.10	0.84	0.18	0.85

3.3. Reactivity Ratio Estimation

Reactivity ratios are estimated by applying the cumulative composition model (using direct numerical integration) to the data through the error-in-variables-model (EVM). The experimental data are presented in Appendix A (Tables A1 and A2), and details regarding the implementation of this technique have been presented previously by Kazemi *et al.* [8].

To better appreciate the error associated with each analysis, reactivity ratio point estimates are presented along with their corresponding joint confidence regions (JCRs). JCRs are typically elliptical contours that quantify the level of uncertainty in the parameter estimates; smaller JCRs indicate higher precision and therefore more confidence in the estimation results [7]. The reactivity ratio point estimates for preliminary, T-M-designed and EVM-designed experiments, along with their associated JCRs, are presented in Figure 1. The literature value from McCormick and Chen [26] is also included for comparison purposes.

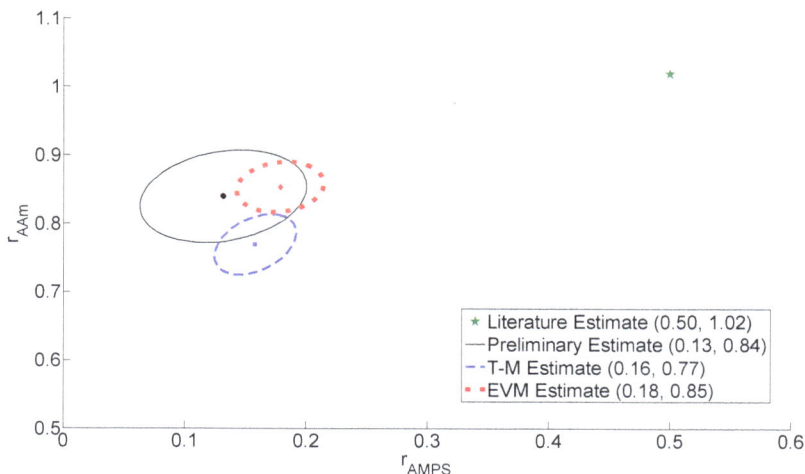

Figure 1. Reactivity ratio estimates for AMPS/AAm copolymer.

Clearly, the reactivity ratio estimates from McCormick and Chen [26] are different from the newly determined reactivity ratios; the estimates from literature are not contained within any of the JCRs. However, it is important to note that the work by McCormick and Chen [26] was at pH = 9, 30 °C and used potassium persulfate (KPS) as the initiator. This is in contrast to the current experimental work, which is at pH = 7, 40 °C and uses ACVA as the initiator. Because the polymerization conditions are different, especially in terms of pH, a difference in results is somewhat expected, although the difference in AMPS values seems considerable.

Overall, all three of the experimental data sets produce similar results. The three JCRs are overlapping, which allows for a high degree of confidence in the results. In comparing the preliminary estimate to the T-M and EVM estimates, the advantages of using optimally designed experiments for reactivity ratio estimation are obvious. The JCRs obtained using the T-M and EVM designs are much smaller than the preliminary design, which indicates that a greater degree of confidence is achieved with the same amount of experimental data.

3.4. Discussion of Results

Reactivity ratios are extremely important parameters in copolymerization kinetics. While the point estimates from the T-M and EVM designs were fairly close, it is still important to establish whether differences in reactivity ratio estimates for the same system will affect subsequent calculations. Reactivity ratios can be used to predict polymer properties such as copolymer composition. Since this information could be used for custom polymer production for specific applications [11], the estimates should be as accurate as possible.

29

3.4.1. Cumulative Copolymer Composition

As an example, the initial feed compositions selected using Tidwell–Mortimer designs are examined in Figure 2. Given the reactivity ratios from the two optimal designs and from literature (see Table 2) and the initial feed compositions ($f_{AMPS,0}$ = 0.30 and $f_{AMPS,0}$ = 0.91), it is possible to predict the cumulative copolymer composition.

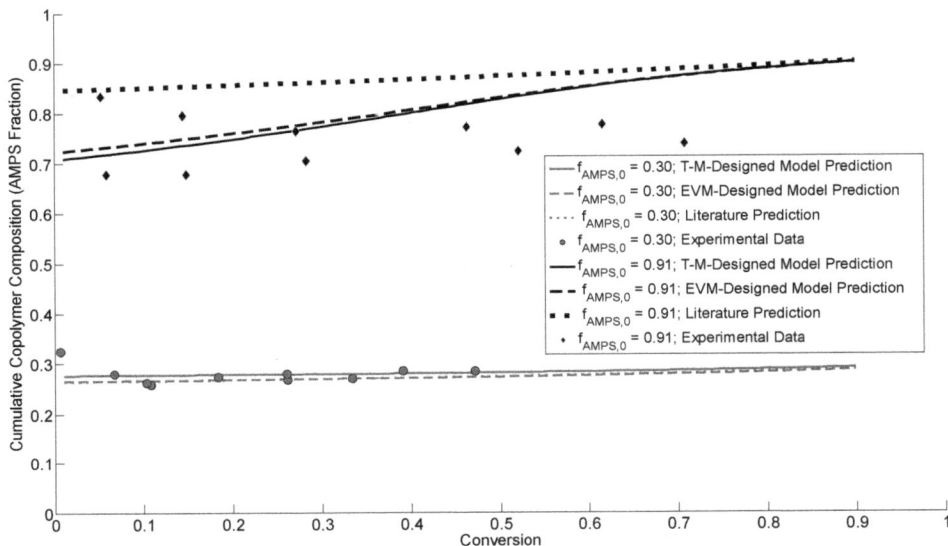

Figure 2. Cumulative copolymer composition for AMPS/AAm; T-M-designed experiments ($f_{AMPS,0}$ = 0.30 and $f_{AMPS,0}$ = 0.91).

This analysis indicates that slight differences in reactivity ratio estimates can significantly affect the cumulative copolymer composition prediction. When the AMPS content in the feed compositions is low (at $f_{AMPS,0}$ = 0.30, for example), the model predictions are in very good agreement. In fact, the model predictions for $f_{AMPS,0}$ = 0.30 from the EVM-design and from McCormick and Chen [26] are almost indistinguishable. However, at $f_{AMPS,0}$ = 0.91, there is a significant difference in model predictions, especially when comparing the optimally-designed experiments to the literature values. The difference in prediction behavior between $f_{AMPS,0}$ = 0.30 and $f_{AMPS,0}$ = 0.91 is due to the nature of the system. When the AMPS content is low in the feed, there is very little composition drift (that is, $f_{AMPS} \approx F_{AMPS}$), which means that the reactivity ratios do not have a significant influence on the copolymer composition predictions. Conversely, when $f_{AMPS,0}$ is high, the propagation of error is evident in the model predictions. Again, this highlights the importance of obtaining accurate reactivity ratios in order to calculate other copolymer property trajectories properly.

3.4.2. Instantaneous Copolymer Composition

The instantaneous copolymer composition can be predicted in the same way that the cumulative copolymer composition was established (using feed compositions and reactivity ratio estimates). As an example, the cumulative and instantaneous composition predictions for $f_{AMPS,0} = 0.84$ are presented in Figure 3.

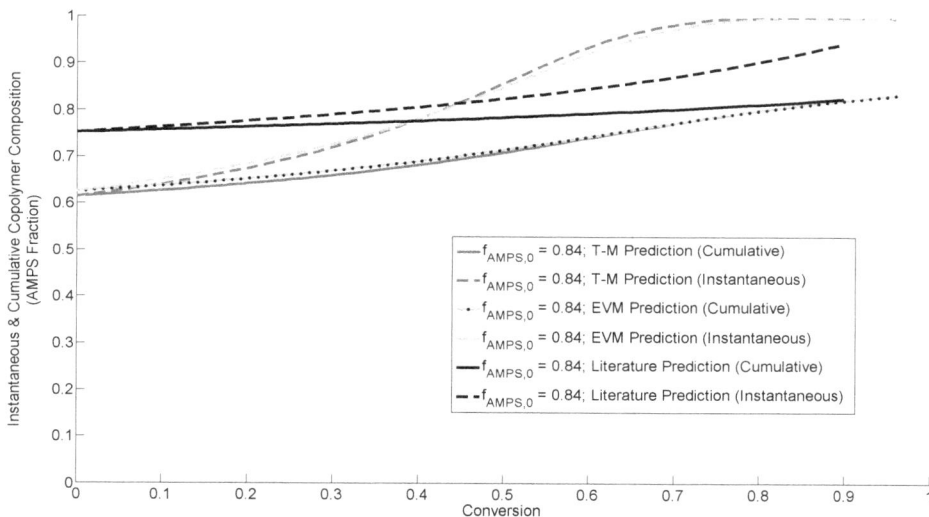

Figure 3. Instantaneous and cumulative copolymer composition predictions (AMPS/AAm).

Figure 3 shows clear similarities between the T-M-designed prediction and the EVM-designed prediction, with slight discrepancies at low conversion. However, the trends are consistent, and the two models seem to converge at higher levels of conversion (>50%). Conversely, the instantaneous and cumulative copolymer composition models using the reactivity ratios from McCormick and Chen [26] give very different results. The initial copolymer composition is at least 10% higher than that predicted by the current investigation, and the trends differ significantly. This is another indication that the reliability of reactivity ratios is extremely important (especially when $f_{AMPS,0}$ is high), which confirms previous observations.

31

3.4.3. Azeotrope Analysis

As mentioned previously, reactivity ratios can be used to estimate the azeotropic composition for a copolymer. Equation (4) can be used to establish azeotropic composition, or F_{AMPS} can be plotted as a function of f_{AMPS} to establish the azeotropic point.

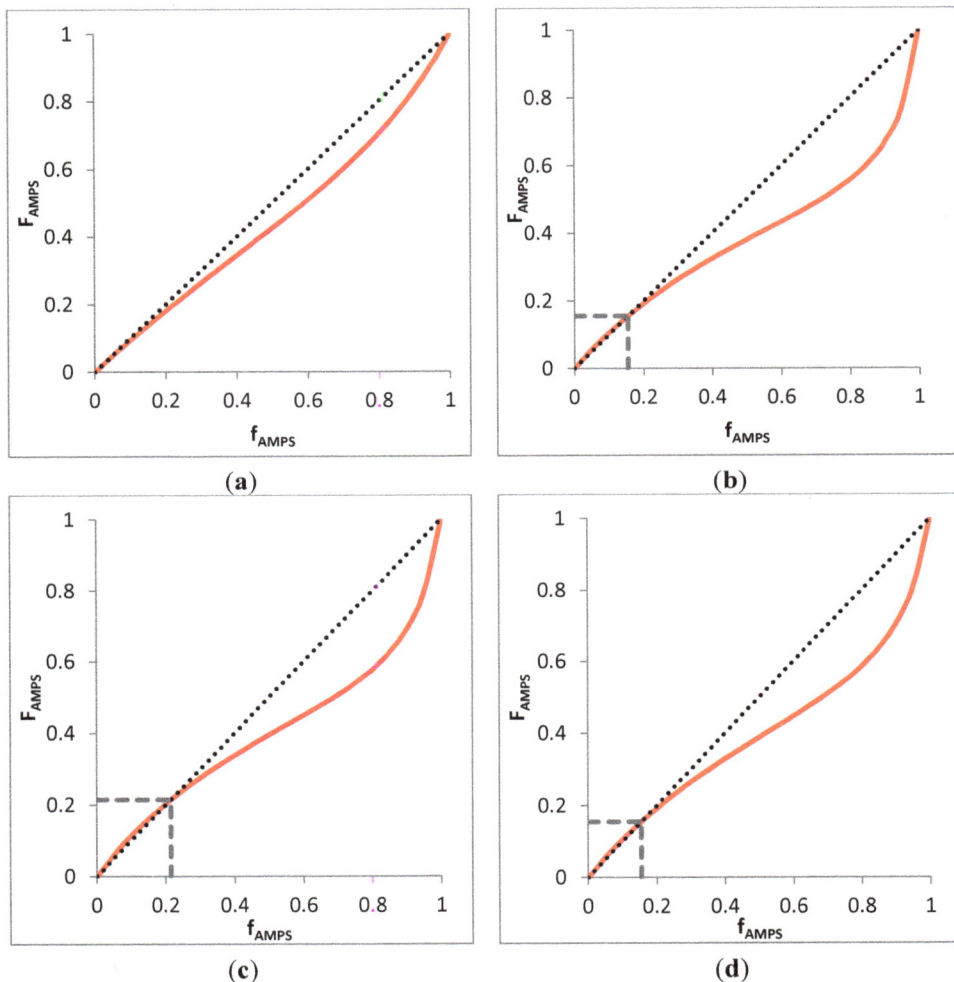

Figure 4. Determination of azeotropic composition from (**a**) literature data [26]; (**b**) preliminary data; (**c**) T-M-designed data; and (**d**) EVM-designed data; 45° line (F_{AMPS} = f_{AMPS}) indicated by a dotted line.

Figure 4 demonstrates how F_{AMPS} varies with f_{AMPS}, given four sets of reactivity ratio estimates ((a) literature data; (b) preliminary data; (c) T-M-designed data; and (d) EVM-designed data). The point at which the curve passes through the 45° line ($F_{AMPS} = f_{AMPS}$, here indicated by a dotted line) represents the azeotropic composition.

It is important to note that $f_{AMPS} \approx F_{AMPS}$ (that is, the F_{AMPS} curve falls very close to the 45° line) at low values of f_{AMPS} in all cases, which confirms the results of Figure 2. However, as expected, the curve never passes through the 45° line in case (a). From a mathematical perspective, it is only feasible to observe a non-negative azeotropic point in the binary system when both reactivity ratios are less than or greater than unity; according to McCormick and Chen [26], $r_{AMPS} = 0.50$ and $r_{AAm} = 1.02$. Therefore, their reactivity ratio estimates suggest that an azeotrope does not occur in this system.

However, in cases (b) through (d), both reactivity ratios are less than unity. Therefore, we expect to observe an azeotrope in the system, and the plots confirm these expectations. Using the reactivity ratios found with T-M-designed data (case (c); $r_{AMPS} = 0.16$ and $r_{AAm} = 0.77$), the azeotrope occurs at $f_{AMPS} = F_{AMPS} = 0.22$. On the other hand, reactivity ratios from the preliminary (case (b); $r_{AMPS} = 0.13$ and $r_{AAm} = 0.84$) and the EVM-designed data (case (d); $r_{AMPS} = 0.18$ and $r_{AAm} = 0.85$) both predict the azeotropic composition to be $f_{AMPS} = F_{AMPS} = 0.16$. Hence, the location of the system azeotrope is somewhere between 0.16 and 0.22.

The agreement between the preliminary and EVM-designed results, combined with the small JCR and high degree of confidence associated with the EVM-designed experiments (even with a limited number of data points, as is evident from Appendix A, give reason to believe that the reactivity ratios obtained through EVM-designed experiments are more trustworthy overall. Therefore, for the AMPS/AAm copolymer system, $r_{AMPS} = 0.18$ and $r_{AAm} = 0.85$.

4. AMPS/AAc Copolymer

4.1. Literature Background for AMPS/AAc

Very few studies have been found with regards to the copolymerization of 2-acrylamido-2-methylpropane sulfonic acid (AMPS) and acrylic acid (AAc). Even fewer have investigated the polymerization kinetics and, specifically, copolymer reactivity ratios. In previous studies, AMPS and AAc have been copolymerized in the presence of crosslinking agents [28–30], and the crosslinked products have been grafted onto backbones via free radical graft polymerization [31,32] to produce hydrogels.

Only two studies [28,33] have been identified that provided reactivity ratio estimates for the AMPS/AAc copolymer along with a description of synthesis and characterization methods (see Table 3). In the work by Abdel-Azim *et al.* [28], reactivity ratios for the AMPS/AAc copolymer were estimated using the Fineman–Ross and Kelen–Tudos (linear) methods. The authors chose to average the two values obtained by the two techniques, which can be a gross approximation for r_{AMPS}.

Table 3. Reactivity ratio estimates for AMPS/AAc copolymer.

Ref.	Experimental	Estimation Technique	r_{AMPS}	r_{AAc}
[28]	–Type: Aqueous solution copolymerization (<10% conversion) –Initiator: BPO –Temperature: 55 °C –Composition: IR	Fineman-Ross Kelen-Tudos Average	0.304 0.15 0.27	0.915 0.98 0.95
[33]	–Type: Aqueous solution copolymerization (<10% conversion) –pH = 7 –Composition: EA	Fineman-Ross Behnken's NLR	0.194 0.187 ± 0.09	0.700 0.740 ± 0.13

Nomenclature: AAc, acrylic acid; AMPS, 2-acrylamido-2-methylpropane sulfonic acid; BPO, benzoyl peroxide; EA, elemental analysis; IR, infrared spectroscopy; NLR, non-linear regression.

4.2. Design of Experiments for AMPS/AAc

Preliminary experiments for AMPS/AAc were based on literature values from Abdel-Azim *et al.* [28]. Both preliminary feed compositions ($f_{AMPS,0} = 0.15$ and $f_{AMPS,0} = 0.80$) presented unique concerns (see Table 4). At the lower AMPS feed composition ($f_{AMPS,0} = 0.15$), the copolymerization was extremely slow and minimal precipitate formed. The high AMPS run ($f_{AMPS,0} = 0.80$) was better in terms of conversion and copolymer precipitation, but presented other difficulties. The reaction took place very quickly, which significantly increased variability in the system. This is, to some extent, characteristic of preliminary experiments, and the error observed in the replicates decreased substantially for the optimally designed experiments.

Fortunately, one of the advantages associated with EVM is the ability to introduce constraints on the experimental design. To avoid the excessively slow polymerization and poor precipitation that was observed for $f_{AMPS,0} = 0.15$, a constraint ($0.2 < f_{AMPS,0} < 1.0$) was included when designing optimal experiments through EVM.

Table 4. Design of experiments and reactivity ratio estimates for AMPS/AAc.

Approach	Reactivity Ratios for Design		Feed Compositions (Mole Fractions)		New Reactivity Ratio Estimates	
	r_{AMPS}	r_{AAc}	$f_{AMPS,0,1}$	$f_{AMPS,0,2}$	r_{AMPS}	r_{AAc}
Preliminary	0.27	0.95	0.15	0.80	0.48	0.95
T-M Design	0.48	0.95	0.32	0.81	0.21	0.85
EVM Design	0.48	0.95	0.20	0.73	0.19	0.86

4.3. Reactivity Ratio Estimation

Reactivity ratios were calculated using conversion and cumulative composition data (see Appendix A, Tables A1 and A2). The cumulative composition model was applied to the data (using direct numerical integration as described in Kazemi *et al.* [8]) through EVM. Point estimates from the literature [28], the preliminary experiments and the optimally designed experiments are presented in Figure 5, along with their corresponding JCRs.

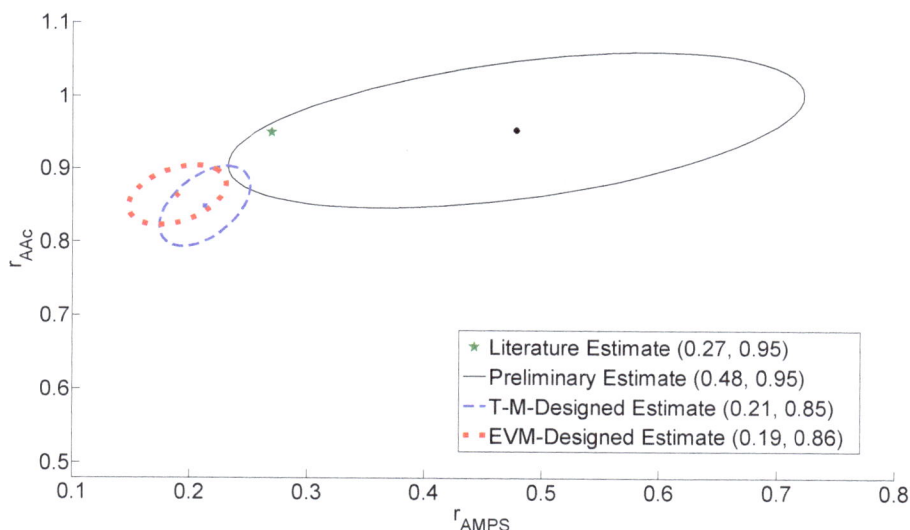

Figure 5. Reactivity ratio estimates for AMPS/AAc copolymer.

The point estimate from Abdel-Azim *et al.* [28] is very close to the edge of the preliminary JCR. While it is reassuring to see that the literature estimates are contained within the preliminary JCR, the preliminary JCR is quite large (as is usually expected for preliminary experimental work). The study by Abdel-Azim *et al.* [28] provided limited insight as to the polymerization conditions for the synthesis of the AMPS/AAc copolymer, but the initiator (benzoyl peroxide) and reaction temperature (55 °C) differed from the currently used conditions. Arguably one of the most important reaction conditions, pH, is not mentioned at all in the work by Abdel-Azim *et al.* [28], so a direct comparison is difficult. However, in general, the estimates made in the previous literature study seem to be close to our newly determined reactivity ratios.

As expected, using experiments that were designed using the Tidwell–Mortimer technique and the error-in-variables-model significantly decreased the error associated with the reactivity ratio estimates. The optimally designed estimates are in relatively good agreement with both the preliminary estimates and the literature values, which allows for a high degree of confidence in the results. The significant overlap between JCRs from the T-M-designed and EVM-designed experiments is also a very good sign, and provides additional confidence in these results.

4.4. Discussion of Results

Because the reactivity ratio estimates from literature, T-M design and EVM design are all similar, it is unlikely that the differences in values will affect composition predictions or other calculations related to copolymer microstructure. However, it is still useful to compare model predictions to experimental results as a confirmation step.

It is reasonable to assume that the EVM-designed results are more accurate, as they were for the AMPS/AAm copolymer (see again Figure 1). However, since the JCRs are close in size (given a similar number of data points in each analysis), it is also helpful to quantify the difference between the two designs. Therefore, the following ratio can be used to compare the confidence regions of the parameters [34]:

$$\frac{\text{Volume}_{\text{T-M Design}}}{\text{Volume}_{\text{EVM Design}}} \propto \left(\frac{|G_{\text{EVM Design}}|}{|G_{\text{T-M Design}}|} \right)^{\frac{1}{2}} \tag{6}$$

where $|G_i|$ is the determinant of the EVM (or T-M) design criterion for a given design of experiments. For the data of Figure 5, the JCR volume ratio is 1.1659, which indicates that the JCR from the T-M design is larger than the JCR from the EVM design [34]. The detailed calculation is provided in Appendix A; the analysis confirms that the EVM-designed experiments produce the smallest JCR for the AMPS/AAc copolymer. An additional advantage of the EVM-designed experiments, which is observed in both Figures 1 and 5, is the decrease in correlation between reactivity ratios compared to the T-M-designed results (as indicated by the decreased slope of the error ellipse). Therefore, the EVM-designed reactivity ratios $r_{\text{AMPS}} = 0.19$ and $r_{\text{AAm}} = 0.86$ can be used to calculate cumulative copolymer composition profiles. Results are shown in Figure 6.

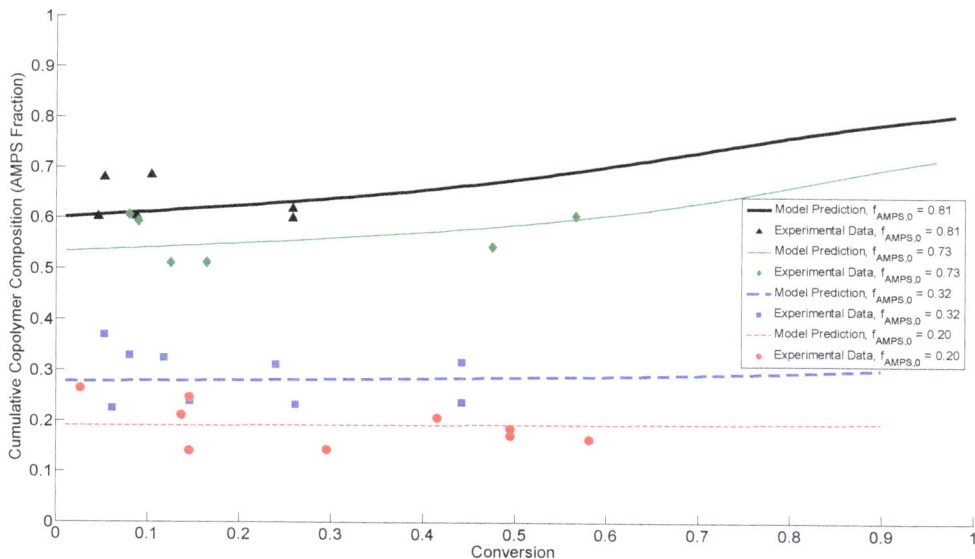

Figure 6. Cumulative copolymer composition for AMPS/AAc.

Here, we see good agreement between the model predictions and the experimental results. Due to the high confidence in reactivity ratio estimates (based on the size of the JCRs and other reasons described earlier), any discrepancies between the data and the model are likely due to errors inherent in the experimental measurements.

5. Conclusions

To improve the shear strength of the AAm/AAc copolymer, often used in enhanced oil recovery, it may be beneficial to add AMPS to the pre-polymerization mixture. To learn more about the AMPS/AAm/AAc terpolymer, reactivity ratios for two associated copolymerizations, AMPS/AAm and AMPS/AAc were established. These binary reactivity ratios can be used with a higher level of confidence (compared to prior literature sources), as many sources of error associated with previous estimation techniques have been removed. The results are summarized in Table 5.

Table 5. Summary of reactivity ratio estimates.

Copolymer	r_1	r_2
AMPS[1]/AAm[2]	0.18	0.85
AMPS[1]/AAc[2]	0.19	0.86

The copolymerization experiments (AMPS/AAm and AMPS/AAc) were designed using two optimal techniques: Tidwell-Mortimer and the error-in-variables-model

(EVM). The best estimates (that is, those with the highest degree of confidence) were those obtained from the EVM-designed data, but both techniques gave similar results. All optimally-designed experiments led to smaller joint confidence regions (JCRs), which is indicative of greater confidence in the reactivity ratio estimates.

Acknowledgments: The authors wish to acknowledge financial support from the Natural Sciences and Engineering Research Council (NSERC) of Canada, and the Canada Research Chair (CRC) program. Thanks also go to UWW/OMNOVA Solutions, USA, for special support to A.J.S.

Author Contributions: This paper is based on the MASc thesis by Alison J. Scott, which was an offspring of the PhD thesis by Marzieh Riahinezhad. Alexander Penlidis supervised both theses.

Conflicts of Interest: The authors declare no conflict of interest.

Appendix A: Experimental Data

Appendix A. A.1. AMPS/AAm Copolymerization Data

Table A1. Experimental data for AMPS/AAm copolymerization; Tidwell-Mortimer design.

Run #	X	$f_{AMPS,0}$	$f_{AAm,0}$	$\bar{F}_{AMPS}$	$\bar{F}_{AAm}$
	0.0061	0.30	0.70	0.3243	0.6757
	0.1078	0.30	0.70	0.2592	0.7408
1	0.2614	0.30	0.70	0.2683	0.7317
	0.3335	0.30	0.70	0.2701	0.7299
	0.4717	0.30	0.70	0.2841	0.7159
	0.0583	0.91	0.09	0.6772	0.3228
	0.1483	0.91	0.09	0.6779	0.3221
2	0.2829	0.91	0.09	0.7043	0.2957
	0.5207	0.91	0.09	0.7223	0.2777
	0.7076	0.91	0.09	0.7374	0.2626
	0.0671	0.30	0.70	0.2794	0.7206
	0.1035	0.30	0.70	0.2626	0.7374
3	0.1830	0.30	0.70	0.2735	0.7265
	0.2604	0.30	0.70	0.2797	0.7203
	0.3910	0.30	0.70	0.2858	0.7142
	0.0519	0.91	0.09	0.8335	0.1665
	0.1441	0.91	0.09	0.7955	0.2045
4	0.2710	0.91	0.09	0.7648	0.2352
	0.4626	0.91	0.09	0.7715	0.2285
	0.6151	0.91	0.09	0.7762	0.2238

X = conversion; $f_{AMPS,0}$ = mole fraction of AMPS in the initial monomer feed; $\bar{F}_{AMPS}$ = cumulative mole fraction (composition) of AMPS in the copolymer product. These symbols are used throughout this Appendix.

Table A2. Experimental data for AMPS/AAm copolymerization; Error-in-Variables-Model design.

Run #	X	$f_{AMPS,0}$	$f_{AAm,0}$	$\overline{F}_{AMPS}$	$\overline{F}_{AAm}$
1	0.3408	0.10	0.90	0.1141	0.8859
	0.3425	0.10	0.90	0.0937	0.9063
	0.7073	0.10	0.90	0.0801	0.9199
2	0.0731	0.84	0.16	0.5977	0.4023
	0.1412	0.84	0.16	0.6332	0.3668
	0.1923	0.84	0.16	0.7141	0.2859
	0.3348	0.84	0.16	0.6555	0.3445
3	0.1064	0.10	0.90	0.1681	0.8319
	0.1473	0.10	0.90	0.0911	0.9089
	0.3556	0.10	0.90	0.0898	0.9102
	0.6174	0.10	0.90	0.0922	0.9078
4	0.2862	0.84	0.16	0.7030	0.2970
	0.3589	0.84	0.16	0.6938	0.3062

Appendix B. A.2. AMPS/AAc Copolymerization Data

Table A1. Experimental data for AMPS/AAc copolymerization; Tidwell-Mortimer design.

Run #	X	$f_{AMPS,0}$	$f_{AAc,0}$	$\overline{F}_{AMPS}$	$\overline{F}_{AAc}$
1	0.0617	0.32	0.68	0.2259	0.7741
	0.1461	0.32	0.68	0.2397	0.7603
	0.2613	0.32	0.68	0.2333	0.7667
	0.4426	0.32	0.68	0.2386	0.7614
	0.4426	0.32	0.68	0.3182	0.6818
2	0.0462	0.81	0.19	0.6014	0.3986
	0.0874	0.81	0.19	0.6032	0.3968
3	0.0528	0.32	0.68	0.3701	0.6299
	0.0804	0.32	0.68	0.3298	0.6702
	0.1177	0.32	0.68	0.3253	0.6747
	0.2395	0.32	0.68	0.3120	0.6880
4	0.0524	0.81	0.19	0.6802	0.3198
	0.1038	0.81	0.19	0.6849	0.3151
	0.2576	0.81	0.19	0.6182	0.3818
	0.2576	0.81	0.19	0.5992	0.4008

Table A2. Experimental data for AMPS/AAc copolymerization; Error-in-Variables-Model design.

Run #	X	$f_{AMPS,0}$	$f_{AAc, 0}$	$\bar{F}_{AMPS}$	$\bar{F}_{AAc}$
1	0.0269	0.20	0.80	0.2652	0.7348
	0.1369	0.20	0.80	0.2119	0.7881
	0.4156	0.20	0.80	0.2075	0.7925
	0.4950	0.20	0.80	0.1860	0.8140
	0.4950	0.20	0.80	0.1723	0.8277
	0.5813	0.20	0.80	0.1649	0.8351
2	0.0895	0.73	0.27	0.5939	0.4061
	0.1250	0.73	0.27	0.5115	0.4885
	0.1642	0.73	0.27	0.5131	0.4869
3	0.1458	0.20	0.80	0.2474	0.7526
	0.1458	0.20	0.80	0.1418	0.8582
	0.2951	0.20	0.80	0.1439	0.8561
4	0.0798	0.73	0.27	0.6063	0.3937
	0.4756	0.73	0.27	0.5455	0.4545
	0.5664	0.73	0.27	0.6069	0.3931

Appendix C. A.3. Design and Joint Confidence Region Comparison Calculations for AMPS/AAc

A common discussion in the field of model-based design of experiments (DOEs) is the need to have a criterion of optimality or efficiency through which designs can be ranked. One comparison metric, the determinant of the information matrix, is related to the volume of the JCR and can therefore determine quantitatively which DOE method is superior. In order to compare two design criteria, a ratio between volumes (obtained from different DOE techniques) is frequently used in the literature [34].

This ratio offers a direct comparison between the sizes of the confidence regions associated with the parameter estimates. Since a more precise parameter estimate (smaller JCR) is indicative of a better and more efficient design, the criterion depends on whether or not the following ratio is greater or less than unity. As shown previously in Equation (6):

$$\frac{Volume_{T-M\ Design}}{Volume_{EVM\ Design}} \propto \left(\frac{|G_{EVM\ Design}|}{|G_{T-M\ Design}|} \right)^{\frac{1}{2}} \quad (A1)$$

where $|G_i|$ is the determinant of the EVM (or T-M) design criterion for a given design of experiments. If the ratio is less than unity, the EVM-designed data has a larger JCR, and the T-M design is superior. Similarly, if the ratio is greater than unity, the T-M-designed data has a larger JCR, and the EVM design is more efficient. The information used for the analysis of AMPS/AAc in Section 4.4 is provided in Table A1, below:

Table A1. Comparison of design criteria for AMPS/AAc copolymerization.

T-M-Designed Data:	EVM-Designed Data:
$G = \begin{bmatrix} 5369.3 & -1869.2 \\ -1869.2 & 2661.0 \end{bmatrix}$ $\|G_{T-M\,Design}\| = 1.0794 \times 10^7$	$G = \begin{bmatrix} 4100.6 & -1608.0 \\ -1608.0 & 4208.8 \end{bmatrix}$ $\|G_{EVM\,Design}\| = 1.4673 \times 10^7$

$$\frac{Volume_{T-M\,Design}}{Volume_{EVM\,Design}} \propto \left(\frac{\|G_{EVM\,Design}\|}{\|G_{T-M\,Design}\|} \right)^{\frac{1}{2}} = \left(\frac{1.4673 \times 10^7}{1.0794 \times 10^7} \right)^{\frac{1}{2}} = 1.1659 \qquad (A2)$$

Since the ratio is greater than unity, the JCR from the T-M design is larger than the JCR from the EVM design [34]. Therefore, the EVM-designed information is more accurate.

References

1. Zaitoun, A.; Makakou, P.; Blin, N.; Al-Maamari, R.; Al-Hashmi, A.; Abdel-Goad, M.; Al-Sharji, H. Shear stability of EOR polymers. In *Society of Petroleum Engineers International Symposium*; Society of Petroleum Engineers: The Woodlands, TX, USA, 2011.
2. Li, Q.; Pu, W.; Wang, Y.; Zhao, T. Synthesis and assessment of a novel AM-co-AMPS polymer for enhanced oil recovery (EOR). In Proceedings of the Fifth International Conference on Computational and Information Sciences, Shiyan, Hubei, China, 21–23 June 2013.
3. Kamal, M.S.; Sultan, A.S.; Al-Mubaiyedh, U.A.; Hussien, I.A.; Pabon, M. Evaluation of rheological and thermal properties of a new fluorocarbon surfactant-polymer system for EOR applications in high-temperature and high-salinity oil reservoirs. *J. Surfactants Deterg.* **2014**, *17*, 985–993.
4. Seright, R.S.; Campbell, A.R.; Mozley, P.S.; Han, P. Stability of partially hydrolyzed polyacrylamides at elevated temperatures in the absence of divalent cations. *SPE J.* **2010**, *15*, 341–348.
5. Riahinezhad, M.; Kazemi, N.; McManus, N.; Penlidis, A. Optimal estimation of reactivity ratios for acrylamide/acrylic acid. *J. Polym. Sci. Part A Polym. Chem.* **2013**, *51*, 4819–4827.
6. Reilly, P.M.; Reilly, H.V.; Keeler, S.E. Parameter estimation in the error-in-variables model. *J. Royal Stat. Soc. Ser. C Appl. Stat.* **1993**, *42*, 693–701.
7. Kazemi, N.; Duever, T.A.; Penlidis, A. A powerful estimation scheme with the error-in-variables model for nonlinear cases: Reactivity ratio estimation examples. *Comput. Chem. Eng.* **2013**, *48*, 200–208.

8.   Kazemi, N.; Duever, T.A.; Penlidis, A. Reactivity ratio estimation from cumulative copolymer composition data. *Macromol. React. Eng.* **2011**, *5*, 385–403.

9.   Brandrup, J.; Immergut, E.H.; Grulke, E.A. *Polymer Handbook*, 4th ed.; Wiley-Interscience: New York, NY, USA, 2003.

10. Kazemi, N.; Duever, T.A.; Penlidis, A. Investigations on azeotropy in multicomponent polymerizations. *Chem. Eng. Technol.* **2010**, *33*, 1841–1849.

11. Riahinezhad, M.; McManus, N.T.; Penlidis, A. Effect of monomer concentration and pH on reaction kinetics and copolymer microstructure of acrylamide/acrylic acid copolymer. *Macromol. React. Eng.* **2015**, *9*, 100–113.

12. Odian, G. *Principles of Polymerization*; Wiley-Interscience: Hoboken, NJ, USA, 2004.

13. Tidwell, P.W.; Mortimer, G.A. An improved method of calculating copolymerization reactivity ratios. *J. Polym. Sci. Part A* **1965**, *3*, 369–387.

14. Kazemi, N.; Duever, T.A.; Penlidis, A. Design of experiments for reactivity ratio estimation in multicomponent polymerizations using the error-in-variables approach. *Macromol. Theory Simul.* **2013**, *22*, 261–272.

15. Riahinezhad, M.; Kazemi, N.; McManus, N.; Penlidis, A. Effect of ionic strength on the reactivity ratios of acrylamide/acrylic acid (sodium acrylate) copolymerization. *J. Appl. Polym. Sci.* **2014**, *131*, 40949.

16. Durmaz, S.; Okay, O. Acrylamide/2-acrylamido-2-methylpropane sulfonic acid sodium salt-based hydrogels: Synthesis and characterization. *Polymer* **2000**, *41*, 3693–3704.

17. Liu, Y.; Xie, J.-J.; Zhu, M.-F.; Zhang, X.-Y. A study of the synthesis and properties of AM/AMPS copolymer as superabsorbent. *Macromol. Mater. Eng.* **2004**, *289*, 1074–1078.

18. Pourjavadi, A.; Salimi, H.; Kurdtabar, M. Hydrolyzed collagen-based hydrogel with salt and pH-responsiveness properties. *J. Appl. Polym. Sci.* **2007**, *106*, 2371–2379.

19. Rosa, F.; Casquilho, M. Effect of synthesis parameters and of temperature of swelling on water absorption by a superabsorbent polymer. *Fuel Process. Technol.* **2012**, *103*, 174–177.

20. Sabhapondit, A.; Borthakur, A.; Haque, I. Characterization of acrylamide polymers for enhanced oil recovery. *J. Appl. Polym. Sci.* **2003**, *87*, 1869–1878.

21. Sabhapondit, A.; Borthakur, A.; Haque, I. Water soluble acrylamidomethyl propane sulfonate. *Energy Fuels* **2003**, *17*, 683–688.

22. Jamshidi, H.; Rabiee, A. Synthesis and characterization of acrylamide-based anionic copolymer and investigation of solution properties. *Adv. Mater. Sci. Eng.* **2014**, *2014*, 1–6.

23. Aggour, Y.A. Thermal degradation of copolymers of 2-acrylamido-2- methylpropanesulphonic acid with acrylamide. *Polym. Degrad. Stab.* **1994**, *44*, 71–73.

24. Billmeyer, F.W. *Textbook of Polymer Science*; John Wiley & Sons, Inc.: New York, NY, USA, 1971.

25. Bune, Y.V.; Barabanova, A.; Bogachev, Y.S.; Gromov, V. Copolymerization of acrylamide with various water-soluble monomers. *Eur. Polym. J.* **1996**, *33*, 1313–1323.

26. McCormick, C.L.; Chen, G.S. Water-soluble copolymers. IV. Random copolymers of acrylamide with sulfonated comonomers. *J. Polym. Sci. Part A Polym. Chem.* **1982**, *20*, 817–838.

27. Travas-Sejdic, J.; Easteal, A. Study of free-radical copolymerization of acrylamide with 2-acrylamido-2-methyl-1-propane sulphonic acid. *J. Appl. Polym. Sci.* **2000**, *75*, 619–628.

28. Abdel-Azim, A.-A.A.; Farahat, M.S.; Atta, A.M.; Abdel-Fattah, A.A. Preparation and properties of two-component hydrogels based on 2-acrylamido-2-methylpropane sulphonic acid. *Polym. Adv. Technol.* **1998**, *9*, 282–289.

29. Liao, L.; Yue, H.; Cui, Y. Crosslink polymerization kinetics and mechanism of hydrogels composed of acrylic acid and 2-acrylamido-2-methylpropane sulfonic acid. *Chin. J. Chem. Eng.* **2011**, *19*, 285–291.

30. Jie, Y.; Pan, Y.; Lu, Q.; Yang, W.; Gao, J.; Li, Y. Synthesis and swelling behaviors of P(AMPS-co-AAc) superabsorbent hydrogel produced by glow-discharge electrolysis plasma. *Plasma Chem. Plasma Process.* **2013**, *33*, 219–235.

31. Pourjavadi, A.; Seidi, F.; Salimi, H.; Soleyman, R. Grafted CMC/Silica gel superabsorbent composite: Synthesis and investigation of swelling behavior in various media. *J. Appl. Polym. Sci.* **2008**, *108*, 3281–3290.

32. Wang, Y.; Shi, X.; Wang, W.; Wang, A. Synthesis, characterization, and swelling behaviors of a pH-responsive CMC-g-poly(AA-co-AMPS) superabsorbent hydrogel. *Turk. J. Chem.* **2013**, *37*, 149–159.

33. Ryles, R.; Neff, R. *Water-Soluble Polymers for Petroleum Recovery*; Stahl, G., Schulz, D., Eds.; Springer: Anaheim, CA, USA, 1986.

34. Kazemi, N. Reactivity Ratio Estimation in Multicomponent Polymerization Systems Using the Error-in-Variables-Model (EVM) Framework. Ph.D. Thesis, Department of Chemical Engineering, University of Waterloo, Waterloo, ON, Canada, 2014.

Modeling of the Copolymerization Kinetics of *n*-Butyl Acrylate and D-Limonene Using PREDICI®

Shanshan Ren, Eduardo Vivaldo-Lima and Marc A. Dubé

Abstract: Kinetic modeling of the bulk copolymerization of D-limonene (Lim) and *n*-butyl acrylate (BA) at 80 °C was performed using PREDICI®. Model predictions of conversion, copolymer composition and average molecular weights are compared to experimental data at five different feed compositions (BA mol fraction = 0.5 to 0.9). The model illustrates the significant effects of degradative chain transfer due to the allylic structure of Lim as well as the intramolecular chain transfer mechanism due to BA.

Reprinted from *Processes*. Cite as: Ren, S.; Vivaldo-Lima, E.; Dubé, M.A. Modeling of the Copolymerization Kinetics of *n*-Butyl Acrylate and D-Limonene Using PREDICI®. *Processes* **2016**, *4*, 1.

1. Introduction

Due to environmental constraints and the need to reduce human dependence on fossil resources, the use of renewable chemical compounds and the incorporation of a naturally-occurring carbon framework into polymer chains has attracted great interest [1,2]. As one of the largest class of renewable feedstocks, terpenes present great potential to replace fossil-based chemical compounds because of their low toxicity, abundant production, and significantly low contribution to the carbon cycle [3–5]. D-limonene (Lim) is a cyclic monoterpene which consists of one isoprene (C_5H_8) unit and is obtained as a by-product from the orange juice industry. For all practical purposes, the free-radical homopolymerization of Lim is not possible. However, the free-radical copolymerization of Lim with various monomers, such as *n*-butyl acrylate (BA) [6], butyl methacrylate (BMA) [7], 2-ethyl hexyl acrylate (EHA) [8], *etc.*, has been reported. In our previous studies [6,9], it was shown that degradative chain transfer due to the presence of Lim competed remarkably with chain propagation. The suppression of both rate of polymerization and molecular weight development were observed. In order to get better insight on the mechanism and the corresponding kinetic parameters related to Lim, a comprehensive model of free-radical copolymerization of BA/Lim was developed using the PREDICI® simulation package. The model is an extension of previous efforts for the BMA/Lim and EHA/Lim systems [10]. The current effort includes further refinement of the Lim rate parameters and addition of an intramolecular chain transfer (*i.e.*, back-biting) mechanism for BA.

BA is a common monomer that is widely used in coating and adhesive formulations due to its excellent resistance to water, solvent and sunlight, as well as the transparency and low-temperature flexibility of its polymer. The mechanism and kinetic parameters of BA have been well-studied for various homo- and copolymerization systems. Recently, it has been reported that the polymerization rate of BA measured by the pulsed-laser polymerization (PLP) method is much slower than expected for chain-end propagation, and this is due to the intramolecular chain-transfer of BA (also referred to as backbiting) yields tertiary radicals which present much slower propagation rates than the typical secondary radicals resulting from chain-end propagation [11–13]. The backbiting mechanism was considered in this work, and the corresponding parameters were mainly taken from Hutchinson and Rantow's work [11,14–17]. Other basic kinetic parameters used in this work were obtained from the WATPOLY database from the University of Waterloo [18–20], which contains parameters for a wide range of monomers, initiators, solvents, CTAs, *etc.*, and can provide good predictions on polymerization rate, composition, and molecular weight in bulk/solution/emulsion systems under a broad range of reaction conditions.

2. Experimental Section

The polymerization conditions and experimental data used herein are from a previous experimental study of the BA/Lim system [6]. Several bulk copolymerizations for five separate BA/Lim feed concentrations were conducted at 80 °C using benzoyl peroxide (BPO) (Sigma-Aldrich, Oakville, ON, Canada) as the initiator. Polymerizations were performed in glass ampoules in an oil bath. Oxygen was removed using several freeze-pump-thaw cycles. Monomer conversion was determined by gravimetry; copolymer composition was measured by ^{1}H-NMR spectroscopy (400 MHz, Bruker Avance, Billerica, MA, USA); and average molecular weights and distribution were obtained by gel permeation chromatography (GPC) (Agilent/Wyatt Technology, Santa Clara/Santa Barbara, CA, USA) equipped with a multi-angle light scattering detector, a differential refractive index detector and a differential viscometer. The initial monomer and initiator concentrations are shown in Table 1.

45

Table 1. Experimental conditions for BA/Lim bulk polymerization at 80 °C.

Monomer Feed (BA Molar Fraction)	BA (mol L^{-1})	Lim (mol L^{-1})	BPO (mol L^{-1})
$f_{BA} = 0.9$	6.13	0.68	0.036
$f_{BA} = 0.8$	5.38	1.35	0.036
$f_{BA} = 0.7$	4.65	2.00	0.036
$f_{BA} = 0.6$	3.91	2.67	0.036
$f_{BA} = 0.5$	3.26	3.26	0.036

3. Model Development

The polymerization mechanism was implemented in PREDICI® and was based on conventional free-radical bulk copolymerization kinetics. Equations describing intramolecular chain transfer (BA) and degradative chain transfer (Lim) were added to the mechanism. Parameter values were taken initially from the literature (including from our previous modeling work on BMA/Lim [10]). The only parameters adjusted were the homopropagation rate constants for BA and Lim. The model equations and initial and final parameter values are shown in Table 2. In Table 2, unreferenced initial parameters were initial guesses for the parameters.

Table 2. Polymerization mechanism and kinetic rate constants used for the PREDICI® simulation of the bulk free-radical copolymerization of BA/Lim at 80 °C [a].

Description	Step in PREDICI®	Variables	Initial Value (L mol^{-1} s^{-1} Unless Otherwise Stated)
	Initiation		
Initiator decomposition	I→2f R$^{\bullet}$	k_d, f	$k_d = 2.52 \times 10^{-5}$ s^{-1} [21], $f = 0.6$
First propagation to Lim	R$^{\bullet}$ + M$_1$→Lim$^{\bullet}$	k_{i1}	1.3 [10], adjusted value = 0.325
First propagation to BA	R$^{\bullet}$ + BA→BA$^{\bullet}$	k_{i2}	4.97×10^4 [17,22]
	Propagation		
Self-propagation of Lim	PLim(s)$^{\bullet}$ + Lim→PLim (s + 1)$^{\bullet}$	k_{p11}	1.3 [10], adjusted value = 0.325
Cross-propagation	PLim(s)$^{\bullet}$ + BA→PBA (s + 1)$^{\bullet}$	k_{p12}	19.9, adjusted value = 48.5
Self-propagation of BA	PBA(s)$^{\bullet}$ + BA→PBA (s + 1)$^{\bullet}$	k_{p22}	4.97×10^4 [17,22], adjusted value = 2.49×10^5
Cross-propagation	PBA(s)$^{\bullet}$ + Lim→PLim (s + 1)$^{\bullet}$	k_{p21}	8.2×10^3, adjusted value = 4.10×10^4
	Chain transfer		
Chain transfer to BA	PLim(s)$^{\bullet}$ + BA→P(s) + BA$^{\bullet}$	k_{fm12}	1.44×10^{-2}
Chain transfer to BA	PBA(s)$^{\bullet}$ + BA→P(s) + BA$^{\bullet}$	k_{fm22}	3.68×10^{-1} [21]
Degradative chain transfer	PLim(s)$^{\bullet}$ + Lim→P(s) + ALim$^{\bullet}$	k_{fm11}	4.65×10^{-1}
Degradative chain transfer	PBA(s)$^{\bullet}$ + Lim→P(s) + ALim$^{\bullet}$	k_{fm21}	2.93×10^2
Re-initiation of ALim$^{\bullet}$	ALim$^{\bullet}$ + Lim→Lim$^{\bullet}$	k_{ra1}	1.67×10^{-10} [10]
Re-initiation of ALim$^{\bullet}$	ALim$^{\bullet}$ + BA→BA$^{\bullet}$	k_{ra2}	1.67×10^{-8} [10]

Table 2. *Cont.*

Description	Step in PREDICI®	Variables	Initial Value (L mol^{-1} s^{-1} Unless Otherwise Stated)
	Termination		
By combination	PLim(s)$^{\bullet}$ + PLim(r)$^{\bullet}$→P(s + r)	k_{tc11}	1.79×10^8
-	PLim(s)$^{\bullet}$ + PBA(r)$^{\bullet}$→P(s + r)	k_{tc12}	1.79×10^8
-	PBA(s)$^{\bullet}$ + PBA(r)$^{\bullet}$→P(s + r)	k_{tc22}	1.79×10^8 [14,17]
-	ALim$^{\bullet}$ + ALim$^{\bullet}$→AA	k_{tcaa}	9.95×10^7
-	PLim(s)$^{\bullet}$ + ALim$^{\bullet}$→P(s)	k_{tc1a}	1.79×10^8
-	PBA(s)$^{\bullet}$ + ALim$^{\bullet}$→P(s)	k_{tc2a}	1.79×10^8
By disproportionation	PLim(s)$^{\bullet}$ + PLim(r)$^{\bullet}$→P(s) + P(r)	k_{td11}	1.99×10^7
-	PLim(s)$^{\bullet}$ + PBA(r)$^{\bullet}$→P(s) + P(r)	k_{td12}	1.99×10^7
-	PBA(s)$^{\bullet}$ + PBA(r)$^{\bullet}$→P(s) + P(r)	k_{td22}	9.95×10^7 [14,17]
	Intramolecular chain transfer of BA		
Backbiting of BA	PBA(s)$^{\bullet}$→QBA(s)$^{\bullet}$	k_{bb}	5.76×10^3 [16,21]
Short-chain branching	QBA(s)$^{\bullet}$+Lim→PLim(s+1)$^{\bullet}$	k_{p21}^{tert}	8.23
Short-chain branching	QBA(s)$^{\bullet}$+BA→PBA(s+1)$^{\bullet}$	k_{p22}^{tert}	49.9 [16,21]
Degradative chain transfer	QBA(s)$^{\bullet}$+Lim→P(s)+ALim$^{\bullet}$	k_{fm2a}^{tert}	1.83
	Termination of BA tertiary radicals		
By combination (QBA$^{\bullet}$)	QBA(s)$^{\bullet}$ + QBA(r)$^{\bullet}$→P(s + r)	$k_{tc22}^{tert-tert}$	1.99×10^7 [14,17]
-	QBA(s)$^{\bullet}$ + PBA(r)$^{\bullet}$→P(s + r)	$k_{tc22}^{tert-sec}$	5.97×10^7 [14,17]
-	QBA(s)$^{\bullet}$ + PLim(r)$^{\bullet}$→P(s + r)	$k_{tc21}^{tert-sec}$	5.97×10^7
-	QBA(s)$^{\bullet}$ + ALim$^{\bullet}$→P(s)	k_{tc2a}^{tert-a}	9.95×10^7
By disproportionation (QBA$^{\bullet}$)	QBA(s)$^{\bullet}$ + QBA(r)$^{\bullet}$→P(s) + P(r)	$k_{td22}^{tert-tert}$	1.79×10^8 [14,17]
-	QBA(s)$^{\bullet}$ + PBA(r)$^{\bullet}$→P(s) + P(r)	$k_{td22}^{tert-sec}$	1.39×10^8 [14,17]
-	QBA(s)$^{\bullet}$ + PLim(r)$^{\bullet}$→P(s) + P(r)	$k_{td21}^{tert-sec}$	1.39×10^8
-	QBA(s)$^{\bullet}$ + ALim$^{\bullet}$→P(s) + A	k_{td2a}^{tert-a}	9.95×10^7

[a] R$^{\bullet}$ = initiator radical; BA and Lim = monomer units; BA$^{\bullet}$ and Lim$^{\bullet}$ = primary radicals; RBA(s)$^{\bullet}$ and RLim(r)$^{\bullet}$= polymer radicals of sizes s and r and ending in BA and Lim, respectively; ALim$^{\bullet}$ = allylic radicals resulting from degradative chain transfer of Lim; QBA$^{\bullet}$ = mid-chain tertiary radical from intramolecular chain transfer of BA; P(s) and P(r) = dead polymers.

3.1. Initiation

The initiation reaction involves two steps:

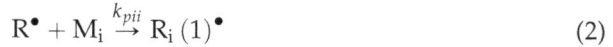

$$I \xrightarrow{k_d} 2f\text{R}^{\bullet} \tag{1}$$

$$\text{R}^{\bullet} + \text{M}_i \xrightarrow{k_{pii}} \text{R}_i(1)^{\bullet} \tag{2}$$

Firstly, the homolysis of BPO initiator (I) yields a pair of primary radicals (R$^{\bullet}$). The decomposition rate of BPO is expressed by an Arrhenius relation, the pre-exponential factor and activation energy values were taken from the WATPOLY simulator database developed by Gao and Penlidis [18–21]:

$$k_d \ (\text{s}^{-1}) = 1.07 \times 10^{14} \exp(-1.515 \times 10^4/T) \tag{3}$$

47

The primary radicals generated by the homolysis step then react with monomer and produce a chain-initiating radical, $R_i(1)^\bullet$. However, not all initiator radicals can react with monomer, they may recombine, or abstract a proton from limonene. These factors were considered in the simulation by introducing an initiator efficiency factor (f). The value of f was set at 0.6, meaning 60% of the primary radicals produced by homolysis could initiate the polymerization.

3.2. Propagation

Using the terminal model, a total of four homo- and cross-propagation reactions were considered:

$$R_i (1)^\bullet + M_j \xrightarrow{k_{pij}} R_j (s+1)^\bullet \tag{4}$$

where k_{pij} is the rate constant of monomer j (M_j) adding to a propagating chain radical ending in monomer i. Note that in this work, 1 refers to limonene (Lim) and 2 refers to n-butyl acrylate (BA). The initially guessed propagation rate constant of BA was:

$$k_{p22} \text{ (L mol}^{-1}\text{s}^{-1}) = 2.21 \times 10^7 \exp(-2153/T) \tag{5}$$

The parameter values were taken from a comprehensive study of BA propagation rate constants using the pulsed-laser polymerization method [23]. The initially-guessed homopolymerization rate constant for Lim (k_{p11}) was taken from our previous modeling study of BMA/Lim copolymerization [10]. The cross-propagation rate constants, k_{p12} and k_{p21} were calculated from the reactivity ratios. Using terminal model kinetics, the reactivity ratios are defined as $r_i = \dfrac{k_{pii}}{k_{pij}}$. The values of $r_1 = 0.0067$ and $r_2 = 6.007$ were determined previously using low conversion bulk experiments at 80 °C [6]. As noted above, the homopropagation rate constants for BA and Lim were the only parameters adjusted in this work. Of course, because of the reactivity ratios, this also resulted in an adjustment to the cross-propagation rate constants.

3.3. Chain Transfer to BA and the Degradative Chain Transfer of Lim

In bulk polymerization, the influence of chain transfer to monomer on molecular weight cannot be ignored due to the high concentration of monomer. The chain transfer to BA is expressed as:

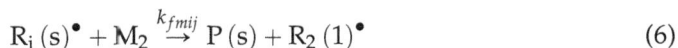

$$R_i (s)^\bullet + M_2 \xrightarrow{k_{fmij}} P (s) + R_2 (1)^\bullet \tag{6}$$

The initially guessed rate constant for chain transfer to BA was taken from the WATPOLY database [21,23]:

$$k_{fm22} \text{ (L mol}^{-1}\text{s}^{-1}) = 1.56 \times 10^4 \exp(-3762/T) \tag{7}$$

As demonstrated in our previous study [6], the highly reactive allylic hydrogen of Lim can easily be abstracted by the growing polymer radical, and yield an inactive chain along with an allylic radical (see Scheme 1). Since the allylic radical is very stable, it is highly unlikely to initiate additional propagation; this mechanism is referred to as degradative chain transfer, and is the dominant chain transfer reaction in the BA/Lim system:

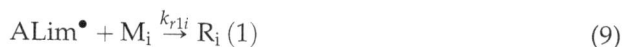

$$R_i (s)^{\bullet} + Lim \overset{k_{fmi1}}{\rightarrow} P (s) + A_{Lim}^{\bullet} \tag{8}$$

$$ALim^{\bullet} + M_i \overset{k_{r1i}}{\rightarrow} R_i (1) \tag{9}$$

Here the symbol $ALim^{\bullet}$ was used to distinguish the allylic radical from the propagating Lim radical ($RLim^{\bullet}$). To obtain an estimate of the chain transfer constant to Lim (C_s), the Mayo equation was used:

$$\frac{1}{\overline{X}_n} = \frac{1}{\overline{X}_{n0}} + C_s \frac{[S]}{[M]} \tag{10}$$

where C_s is defined as the ratio of the chain transfer to Lim rate constant to the BA propagation rate constant; that is, $C_s = \dfrac{k_{fm21}}{k_{p22}}$, $\overline{X}_n$ is the number-average degree of polymerization, $\overline{X}_{n0}$ is the number-average degree of polymerization in the absence of solvent/chain transfer agent, $[S]$ and $[M]$ are the molar concentrations of solvent/chain transfer agent and/or monomer, respectively. By plotting $\dfrac{1}{\overline{X}_n}$ vs. $\dfrac{[S]}{[M]}$, C_s was calculated from the slope as 4.9×10^{-3}. An assumption that Lim acts more like a solvent or chain transfer agent rather than a propagating monomer was used to simplify the equation. The corresponding Mayo plot is given in Figure 1.

To simplify the model, it is reasonable to assume the propagating radicals present the same chain transfer reactivity (H-atom abstraction) to a particular monomer as their propensity of adding to that monomer during propagation [11]. Accordingly, k_{fm21} and k_{fm12} were calculated as:

$$\frac{k_{fm21}}{k_{fm11}} = \frac{k_{p21}}{k_{p11}} \text{ and } \frac{k_{fm12}}{k_{fm22}} = \frac{k_{p12}}{k_{p22}} \tag{11}$$

49

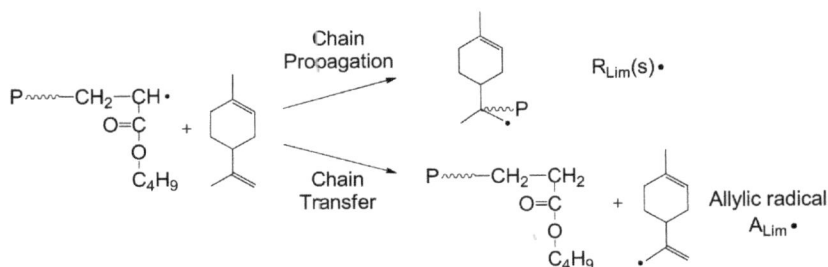

Scheme 1. Ideal reaction schematic of chain propagation and degradative chain transfer of Lim.

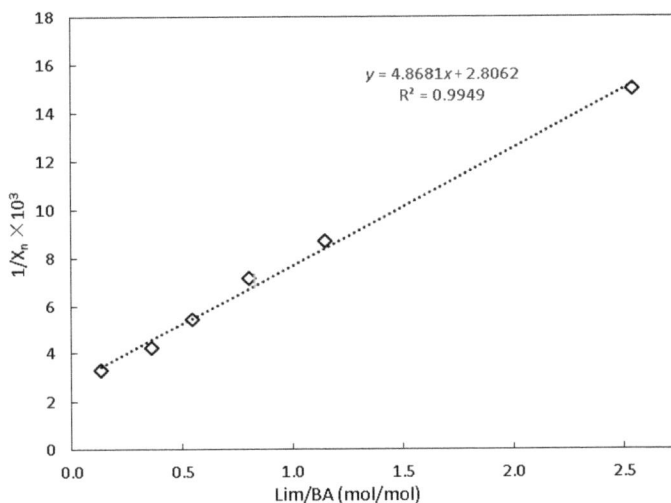

Figure 1. Mayo plot for the estimation of chain transfer constant to Lim (C_s).

The value of the re-initiation rate constant for the allylic radical was assumed to be very small as the radical is stable and would not be expected to re-initiate a new propagating chain. Values of 1.67×10^{-10} L mol^{-1} s^{-1} and 1.67×10^{-8} L mol^{-1} s^{-1} were used for k_{r11} and k_{r12}, respectively, according to a previous study [10].

3.4. Termination

The termination reaction occurs by both combination and disproportionation of polymer radicals (see also Scheme 2):

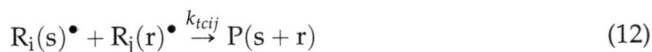

$$R_i(s)^\bullet + R_j(r)^\bullet \overset{k_{tcij}}{\rightarrow} P(s+r) \tag{12}$$

$$R_i(s)^\bullet + R_j(r)^\bullet \overset{k_{tdij}}{\rightarrow} P(s) + P(r) \tag{13}$$

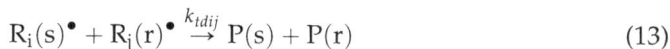

The overall termination rate constant of BA, $k_{t22} = k_{tc22} + k_{td22}$, was fitted to an Arrhenius expression as [17,24]:

$$k_{t22} \ (\text{L mol}^{-1}\text{s}^{-1}) = 1.34 \times 10^9 \ \exp(-674/T) \tag{14}$$

The termination rate constants of Lim-related species, *i.e.*, k_{tc11}, k_{td11}, k_{tcaa}, k_{tc1a}, and k_{tc2a}, and cross-termination, *i.e.*, k_{tc12} and k_{td12}, were set to the same level as for BA radicals. Due to the lack of any known values for these parameters in the literature and because BA feed concentrations were at 50% or higher, this was considered a best option for an initial guess. The ratio of termination by combination to overall termination rate ($\frac{k_{tc}}{k_t}$), is taken as 0.9 for both the BA and Lim radicals as recommend by Peck and Hutchinson for the termination for secondary-secondary radicals [14].

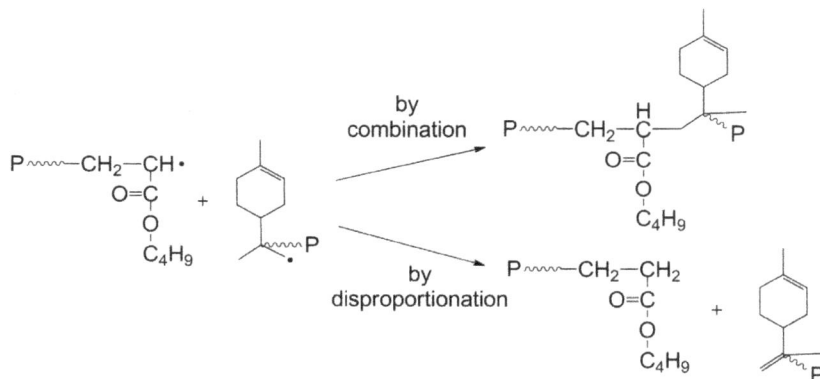

Scheme 2. Ideal reaction schematic of termination reaction of BA/Lim copolymerization.

3.5. Backbiting of BA

There is significant evidence that intramolecular chain transfer to polymer is significant during the chain propagation reaction of BA [14,25,26]. The mid-chain tertiary radicals resulting from a backbiting mechanism are quite stable and present slower propagation rates compared to the secondary radicals resulting from regular chain-end propagation. The propagation of tertiary radicals creates short-chain branches in the polymer. The backbiting and short-chain branching mechanisms were included in this model:

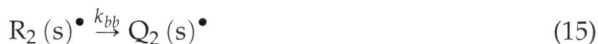

$$R_2 \ (s)^\bullet \overset{k_{bb}}{\rightarrow} Q_2 \ (s)^\bullet \tag{15}$$

$$Q_2 \, (s)^\bullet + M_i \xrightarrow{k_{p2i}{}^{tert}} R_i \, (s+1)^\bullet \tag{16}$$

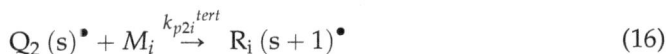

The symbol $Q^\bullet$ represents the mid-chain tertiary radical. The rate constants for backbiting (k_{bb}) and short branching propagation ($k_{p2i}{}^{tert}$) were fitted using [16,21]:

$$k_{bb} \, (\text{L mol}^{-1}\text{s}^{-1}) = 3.87 \times 10^6 \exp(-2299/T) \tag{17}$$

$$k_{p22}{}^{tert} \, (\text{L mol}^{-1}\text{s}^{-1}) = 59.9 \exp(-64.2/T) \tag{18}$$

$$k_{p21}{}^{tert} = \frac{k_{p22}{}^{tert}}{r_2} \tag{19}$$

The mid-chain tertiary radical can terminate with either a tertiary radical or a chain-end secondary radical:

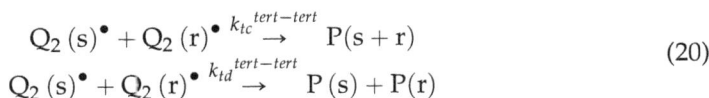

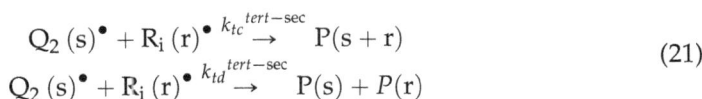

$$\begin{aligned} Q_2 \, (s)^\bullet + Q_2 \, (r)^\bullet \xrightarrow{k_{tc}{}^{tert-tert}} P(s+r) \\ Q_2 \, (s)^\bullet + Q_2 \, (r)^\bullet \xrightarrow{k_{td}{}^{tert-tert}} P(s) + P(r) \end{aligned} \tag{20}$$

$$\begin{aligned} Q_2 \, (s)^\bullet + R_i \, (r)^\bullet \xrightarrow{k_{tc}{}^{tert-sec}} P(s+r) \\ Q_2 \, (s)^\bullet + R_i \, (r)^\bullet \xrightarrow{k_{td}{}^{tert-sec}} P(s) + P(r) \end{aligned} \tag{21}$$

By assuming the termination rate constant for BA is independent of radical type, the overall termination rate constant of tertiary radicals is taken to be the same as k_{t22} [14,17]. Unlike secondary radicals, termination by disproportionation is favored by tertiary radicals. The ratio of termination by combination to overall termination rate was taken as 0.1 for tertiary-tertiary radicals, 0.3 for tertiary-secondary radicals [14,21], and 0.5 for tertiary-allylic radicals.

All the rate constants were assumed as chain-length independent for the purposes of model simplification. Note that diffusion-control effects were not considered in the BA/Lim kinetic model, since the polymerization temperature is much higher than the glass transition temperature of the mixture, so that the molecules are in a rubbery (mobile) state. In the BA/Lim data used in this study, conversion was kept relatively low (due to the degradative chain transfer presented by Lim) and the reaction medium was not viscous enough to induce diffusion-controlled behavior.

4. Results And Discussion

4.1. Backbiting of BA

Figure 2 shows conversion *vs.* time data and Figure 3 shows molecular weight *vs.* time data along with model predictions at $f_{BA} = 0.9$ with and without considering the backbiting mechanism using the initial rate constants. As the data illustrate, modeling without considering backbiting led to an overestimation of conversion results, while modeling with backbiting showed a considerable underestimation of the results. This reflects the fact that the backbiting mechanism yields tertiary radicals which exhibit a much lower reactivity, and the overall reaction rate is therefore reduced. In Figure 3, the model with backbiting provides a more reasonable prediction of the average molecular weights, whereas the model without backbiting yielded a much higher prediction. Each backbiting event results in the creation of a short branch on the main polymer chain, and the molecular weight is decreased accordingly. In order to balance the conversion and molecular weight simulation results, the rate constants of propagation for both species (BA and Lim) were adjusted to new values to present better predictions (see values in Table 2). It is important to point out that k_p for BA was estimated using the experimental data at high BA content ($f_{BA} = 0.9$), whereas k_p for Lim was estimated using the experimental data at low BA content ($f_{BA} = 0.5$).

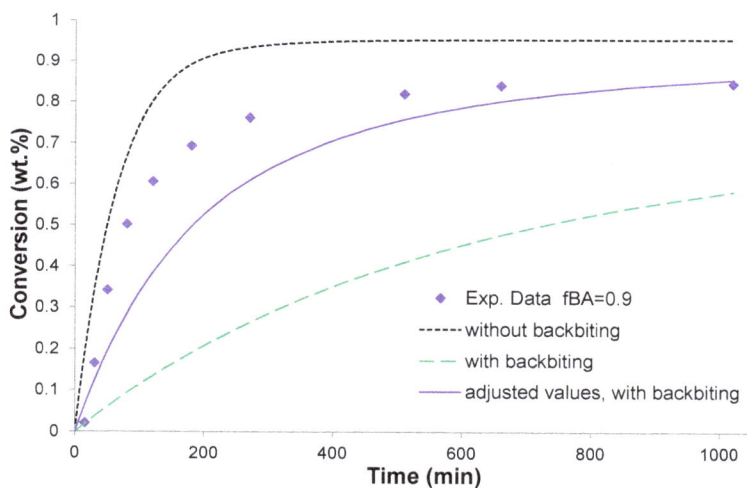

Figure 2. Conversion *vs.* time simulation with and without backbiting, BA/Lim copolymerization at feed composition $f_{BA} = 0.9$, at 80 °C in bulk using BPO (1 wt.%).

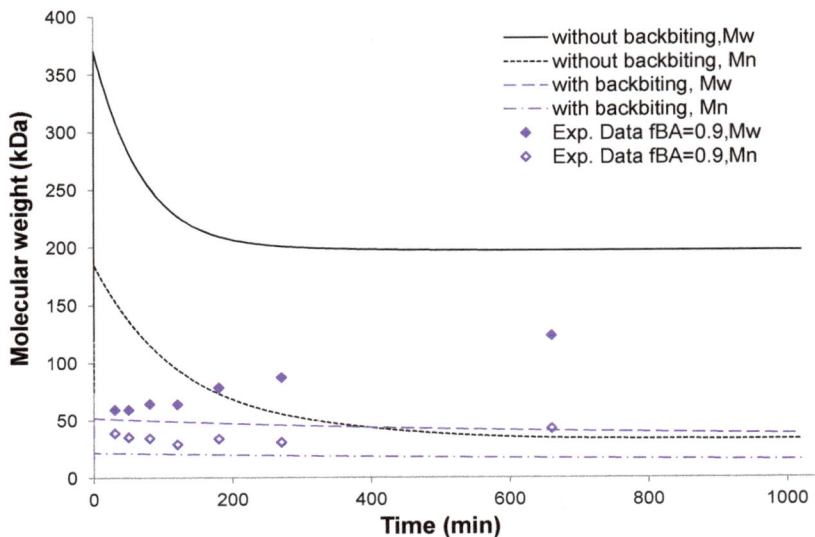

Figure 3. Molecular weight $vs.$ time simulation with and without backbiting, BA/Lim copolymerization at feed composition $f_{BA} = 0.9$, at 80 °C in bulk using BPO (1 wt.%).

4.2. Conversion vs. Time Results

The conversion $vs.$ time model predictions at five different initial feed compositions (f_{BA} = 0.5 to 0.9) along with the experimental data are shown in Figure 4. The agreement between the model and the experimental data is reasonably good. The model prediction trends in the data are well-predicted by the model; increases in BA feed content resulted in higher reaction rates. However, predictions at higher BA feed fractions were less impressive. One possible explanation could be that the degradative chain transfer reaction of Lim competes with the backbiting mechanism, as the BA chain-end radicals possibly abstracted the hydrogen from the Lim molecule rather than from an acrylate unit on its own chain. In other words, backbiting, which is the main cause for a decrease in the polymerization rate, is less dominant in the presence of Lim.

4.3. Copolymer Composition vs. Conversion

Composition $vs.$ conversion profiles for different feed compositions are shown in Figure 5. As mentioned earlier, the propagation rate constants were calculated using the reactivity ratios previously estimated from low-conversion BA/Lim experiments [6]. The agreement between model profiles and experimental data are in general, very good. The good predictions of the composition $vs.$ conversion data validates the reactivity ratio values estimated.

54

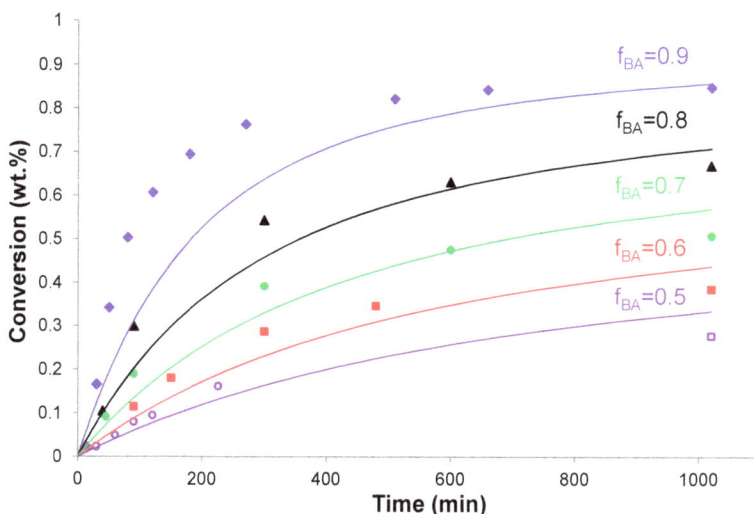

Figure 4. Conversion *vs.* time profiles for BA/Lim bulk copolymerizations at various feed compositions. Solid lines are model predictions.

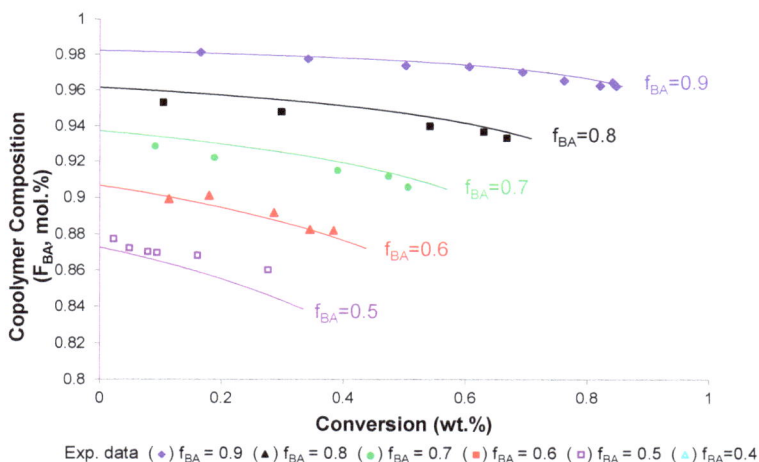

Figure 5. Composition *vs.* conversion profiles for BA/Lim copolymerizations at various feed compositions. Solid lines are model prediction.

4.4. Molecular Weight of Soluble Copolymer

Figure 6 shows plots of the molecular weight development *vs.* conversion at four different feed compositions. The number-average molecular weight (M_n) is relatively well predicted by the model but the weight-average molecular weight (M_w) shows some discrepancy. Nonetheless, good prediction of molecular weight data is often difficult to achieve. One can, however, make important conclusions about the reaction mechanisms despite some model mismatch. The model reveals that the average molecular weight decreases significantly with increasing Lim concentration in the feed. Given the propagation and chain transfer rate constants for Lim fitted to this model (see Table 2), it can be concluded that Lim acts more like a chain transfer agent than a co-monomer. One may also note that M_n decreases with increasing conversion, which is consistent with the fact that more short-chain polymers were produced. The production of increasing amounts of short-chain polymers likely resulted from increased degradative chain transfer due to increased Lim concentration as BA was preferentially incorporated into the copolymer during the early stages of the polymerization (see reactivity ratios described earlier, and Figure 5).

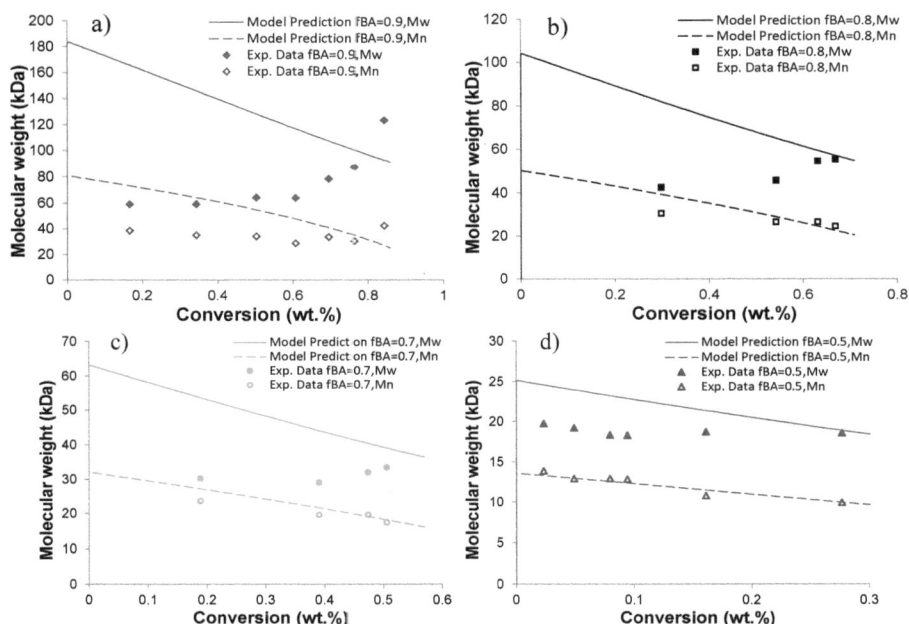

Figure 6. Molecular weight *vs.* conversion profiles for BA/Lim copolymerization at various feed compositions: (**a**) $f_{BA} = 0.9$, (**b**) $f_{BA} = 0.8$, (**c**) $f_{BA} = 0.7$, and (**d**) $f_{BA} = 0.5$.

One possible explanation for the broadening polydispersity evident from Figure 6 could relate to long-chain branching caused by the intermolecular chain transfer of BA (a propagating radical abstracts the hydrogen from an acrylate unit in the middle of another chain) [14], which can result in a significant increase in M_w. In this case, additional studies would be required to pursue these ideas further.

5. Conclusions

A kinetic model of the bulk polymerization of BA/Lim has been developed with the addition of important mechanisms for degradative chain transfer to Lim and backbiting for BA. Rate constants related to BA and Lim species were calculated based on literature values, reactivity ratios and degradative chain transfer constants estimated from previous experimental results. Fitting of the propagation rate constants resulted in moderately good conversion and molecular weight predictions and very good predictions of copolymer composition. The model supports the presence of a significant degradative chain transfer to Lim reaction as well as a backbiting mechanism for BA. Future work including a long-chain branching mechanism may shed further light on this copolymer system. In any event, this work provides greater insight into the use of an important bio-based, renewable monomer.

Acknowledgments: The authors acknowledge the support of the China Scholarship Commission for the support of Ms. Shanshan Ren. Furthermore, financial support for this project by Intellectual Ventures and the Natural Sciences and Engineering Research Council (NSERC) of Canada is greatly appreciated. E. Vivaldo-Lima acknowledges support from DGAPA-UNAM (PASPA Program) and the University of Ottawa Distinguished Visiting Researcher Program.

Author Contributions: Marc A. Dubé and Shanshan Ren conceived the experimental research. Shanshan Ren conducted all experiments and characterization. Eduardo Vivaldo-Lima performed all simulations with feedback from Shanshan Ren. Marc A. Dubé and Eduardo Vivaldo-Lima provided technical direction to the project. Shanshan Ren wrote the paper while Marc A. Dubé and Eduardo Vivaldo-Lima revised the final document.

Conflicts of Interest: The authors declare no conflict of interest.

References

1. Belgacem, M.N.; Alessandro, G. *Monomers, Polymers and Composites from Renewable Resources*; Elsevier: Amsterdam, The Netherlands, 2008; p. 543.
2. Ahmed, J.; Tiwari, B.K.; Imam, S.H.; Rao, M.A. *Starch-based Polymeric Materials and Nanocomposites: Chemistry, Processing, and Applications*; CRC Press: Boca Raton, FL, USA, 2012.
3. Kim, Y.W.; Kim, M.J.; Chung, B.Y.; Bang, D.Y.; Lim, S.K.; Choi, S.M.; Lim, D.S.; Cho, M.C.; Yoon, K.; Kim, H.S.; *et al.* Safety evaluation and risk assessment of D-limonene. *J. Toxicol. Environ. Health Pt B* **2013**, *16*, 17–38.
4. Ciriminna, R.; Lomeli-Rodriguez, M.; Demma Cara, P.; Lopez-Sanchez, J.; Pagliaro, M. Limonene: A versatile chemical of the bioeconomy. *Chem. Commun.* **2014**, *50*, 15288–15296.
5. Wilbon, P.A.; Chu, F.; Tang, C. Progress in renewable polymers from natural terpenes, terpenoids, and rosin. *Macromol. Rapid Commun.* **2013**, *34*, 8–37.
6. Ren, S.; Trevino, E.; Dubé, M.A. Copolymerization of limonene with *n*-butyl acrylate. *Macromol. React. Eng.* **2015**, *9*, 339–349.

7. Zhang, Y.; Dubé, M.A. Copolymerization of *n*-butyl methacrylate and D-limonene. *Macromol. React. Eng.* **2014**, *8*, 805–812.

8. Zhang, Y.; Dubé, M.A. Copolymerization of 2-Ethylhexyl acrylate and D-limonene. *Polym. Plast Technol. Eng* **2015**, *54*, 499–505.

9. Ren, S.; Zhang, L.; Dubé, M.A. Free-radical terpolymerization of *n*-butyl acrylate/butyl methacrylate/D-limonene. *J. Appl. Polym. Sci.* **2015**, *132*, 42821–42829.

10. Zhang, Y.; Dubé, M.A.; Vivaldo-Lima, E. Modeling degradative chain transfer in D-Limonene/*n*-butyl methacrylate free-radical copolymerization. *J. Renew. Mat.* **2015**, *3*, 318–326.

11. Li, D.; Grady, M.C.; Hutchinson, R.A. High-temperature semibatch free radical copolymerization of butyl methacrylate and butyl acrylate. *Ind. Eng. Chem. Res.* **2005**, *44*, 2506–2517.

12. Beuermann, S.; Paquet, D.A.; McMinn, J.H.; Hutchinson, R.A. Determination of free-radical propagation rate coefficients of butyl, 2-ethylhexyl, and dodecyl acrylates by pulsed-laser polymerization. *Macromolecules* **1996**, *29*, 4206–4215.

13. Lyons, R.A.; Hutovic, J.; Piton, M.C.; Christie, D.I.; Clay, P.A.; Manders, B.G.; Kable, S.H.; Gilbert, R.G. Pulsed-laser polymerization measurements of the propagation rate coefficient for butyl acrylate. *Macromolecules* **1996**, *29*, 1918–1927.

14. Peck, A.N.F.; Hutchinson, R.A. Secondary reactions in the high-temperature free radical polymerization of butyl acrylate. *Macromolecules* **2004**, *37*, 5944–5951.

15. Nikitin, A.N.; Hutchinson, R.A. Effect of intramolecular transfer to polymer on stationary free radical polymerization of alkyl acrylates, 2. *Macromol. Theory Simul.* **2006**, *15*, 128–136.

16. Rantow, F.S.; Soroush, M.; Grady, M.C.; Kalfas, G.A. Spontaneous polymerization and chain microstructure evolution in high-temperature solution polymerization of *n*-butyl acrylate. *Polymer* **2006**, *47*, 1423–1435.

17. Nikitin, A.N.; Hutchinson, R.A.; Buback, M.; Hesse, P. Determination of intramolecular chain transfer and midchain radical propagation rate coefficients for butyl acrylate by pulsed laser polymerization. *Macromolecules* **2007**, *40*, 8631–8641.

18. Gao, J.; Penlidis, A. A Comprehensive simulator/database package for reviewing free-radical homopolymerizations. *J. Macromol. Sci. Part C* **1996**, *36*, 199–404.

19. Gao, J.; Penlidis, A. A Comprehensive simulator/database package for reviewing free-radical copolymerizations *J. Macromol. Sci. Part C* **1998**, *38*, 651–780.

20. Gao, J.; Penlidis, A. A Comprehensive simulator/database package for bulk/solution free-radical terpolymerizations. *Macromol. Chem. Phys.* **2000**, *201*, 1176–1184.

21. Dorschner, D. Multicomponent Free Radical Polymerization Model Refinements and Extensions with Depropagations. Master's Thesis, University of Waterloo, Department of Chemical Engineering, Waterloo, ON, Canada, 2010.

22. Asua, J.M.; Beuermann, S.; Buback, M.; Castignolles, P.; Charleux, B.; Gilbert, R.G.; Hutchinson, R.A.; Leiza, J.R.; Nikitin, A.N.; Vairon, J.; *et al.* Critically evaluated rate coefficients for free-radical polymerization, 5. *Macromol. Chem. Phys.* **2004**, *205*, 2151–2160.

23. Jun, W. Mathematical Modeling of Free-Radical Six Component Bulk and Solution Polymerization. Master's Thesis, University of Waterloo, Department of Chemical Engineering, Waterloo, ON, Canada, 2008.

24. Beuermann, S.; Buback, M. Rate coefficients of free-radical polymerization deduced from pulsed laser experiments. *Prog. Polym. Sci.* **2002**, *27*, 191–254.

25. Plessis, C.; Arzamendi, G.; Leiza, J.R.; Schoonbrood, H.A.S.; Charmot, D.; Asua, J.M. modeling of seeded semibatch emulsion polymerization of *n*-BA. *Ind. Eng. Chem. Res.* **2001**, *40*, 3883–3894.

26. Ahmad, N.M.; Heatley, F.; Lovell, P.A. Chain transfer to polymer in free-radical solution polymerization of *n*-butyl acrylate studied by NMR spectroscopy. *Macromolecules* **1998**, *31*, 2822–2827.

State Observer Design for Monitoring the Degree of Polymerization in a Series of Melt Polycondensation Reactors

Chen Ling and Costas Kravaris

Abstract: A nonlinear reduced-order state observer is applied to estimate the degree of polymerization in a series of polycondensation reactors. The finishing stage of polyethylene terephthalate synthesis is considered in this work. This process has a special structure of lower block triangular form, which is properly utilized to facilitate the calculation of the state-dependent gain in the observer design. There are two possible on-line measurements in each reactor. One is continuous, and the other is slow-sampled with dead time. For the slow-sampled titration measurement, inter-sample behavior is estimated from an inter-sample output predictor, which is essential in providing continuous corrections on the observer. Dead time compensation is carried out in the same spirit as the Smith predictor to reduce the effect of delay in the measurement outputs. By integrating the continuous-time reduced-order observer, the inter-sample predictor and the dead time compensator together, the degree of polymerization is accurately estimated in all reactors. The observer performance is demonstrated by numerical simulations. In addition, a pre-filtering technique is used in the presence of sensor noise.

Reprinted from *Processes*. Cite as: Ling, C.; Kravaris, C. State Observer Design for Monitoring the Degree of Polymerization in a Series of Melt Polycondensation Reactors. *Processes* **2016**, *4*, 4.

1. Introduction

Polymers are continuously substituting traditional materials (e.g., glass, woods and metals) along with low cost and good processability. Polyethylene terephthalate (PET) is the most common thermoplastic polymer resin, which is the primary raw material for synthetic fibers, dielectric films and beverage bottles. PET has dominated the synthetic fibers industry over the years accounting for nearly half of the global consumption [1]. Moreover, the global demand for PET is predicted to grow in the next few years. Therefore, producing PET with the required properties is of major industrial importance.

It is well known that the end-use properties of PET, such as drawing behavior, melting point, tensile strength and thermal stability, strongly depend on its molecular weight and byproduct concentrations [2–5]. There are several side reactions taking place along with the main polycondensation reaction. The amount of side products (*i.e.*, diethylene glycol (DEG), acetaldehyde, water, carboxyl end groups, vinyl end groups) determines the quality and properties of the final PET product. For example, every one percent of DEG in the polyester chain will cause a lower melting point by 5 °C [6]. Additionally, even a small amount of DEG leads to reduced heat resistance, decreased crystallinity and UV light stability. Vinyl end groups may also be polymerized with other polyester chains to form polyvinyl ester, of which the pyrolysis products have been shown to be responsible for the coloration of PET [7]. A high initial concentration of carboxyl groups could induce a decrease in the degree of polymerization (DP) due to hydrolytic degradation [8]. In order to ensure product quality, the amount of byproducts needs to be well controlled within certain limits.

However, the monitoring and control of polymerization reactors is not an easy task, owing to a lack of fast on-line measurements and the significant nonlinearity of the processes. Very often, critical quantities related to safety, product quality and/or economic performance of a polymerization process cannot be measured on line. Thus, state estimation plays an important role in providing frequent and reliable information of the process, which can be integrated into model-based control, as well. Since the early 1980s, there have been significant efforts in the design and application of state estimators to polymerization reactions, especially in free radical polymerization. The extended Kalman filter (EKF), as an industrially-popular estimator, has been widely used and achieved fairly good performance in many cases [9–16]. In this approach, the design is based on an approximate local linearization of the system along a reference trajectory. Even though EKF has found industrial applications, there have been studies that established its serious difficulties in the presence of strong process nonlinearities [17,18]. An alternative approach for estimation in polymerization processes is state observer design [19–26]. It utilizes the dynamic process model, which captures the evolution of physical and chemical phenomena, and then generates a soft sensor that is able to reconstruct the missing state variables with additional appropriate feedback terms from all of the on-line measurements. For example, Van Dootingh *et al.* [19] developed a nonlinear high-gain observer with adjustable speed of convergence in a styrene polymerization reactor. Compared to EKF, this observer does not only have a theoretical proof of convergence, but also greatly reduces computation time. Tatiraju and Soroush [21,22] implemented a nonlinear reduced-order observer to a homopolymerization reactor. Along with an open-loop observer for the unobservable states, accurate estimates for all states were achieved. Astorga *et al.* [25] used a continuous-discrete observer to estimate monomer composition in an emulsion copolymerization reactor. The proposed observer was validated by comparing the outputs of the observer with off-line gas chromatography results.

Although a significant amount of work has been done in the monitoring and control of free radical polymerization reactors, very few state estimation studies are available for polycondensation reactors. Choi and Khan [27] applied the EKF algorithm to estimate nine state variables in the transesterification stage of PET synthesis. When supplemented by five additional off-line measurements, the overall performance of the state estimator was greatly improved. Appelhaus *et al.* [28] designed an extended observer to estimate concentrations of ethylene glycol (EG) and hydroxyl end groups along with a mass transfer parameter in a batch reactor. In their study, only the reversible polycondensation reaction was considered.

A comprehensive understanding of PET synthesis is essential for effective quality control and optimization of the process. Generally, there are three stages (*i.e.*, transesterification/esterification, pre-polymerization and polycondensation) involved in PET production. For injection or blow molding applications, solid state polymerization needs to be carried out afterwards to obtain a product with DP over 150. In each reactor, side reactions take place simultaneously and directly affect product quality. On-line measurements for byproduct concentrations are usually not available or at relatively low sampling rates [29]. Therefore, based on the fact that available on-line measurements are not always of the same nature, it is necessary to develop estimation/monitoring algorithms that can use all of these different kinds of on-line measurements in a synergistic way to provide valuable information of the process.

In this study, the nonlinear observer design method of exact linearization with eigenvalue assignment [30,31] is applied to a series of three continuous polycondensation reactors. A modified reaction-mass transfer model [32] is used in our work. The objective is to estimate unmeasured concentrations, as well as the degree of polymerization in the PET finishing stage from continuous hydroxyl measurement and sampled acidimetric titration, where different sampling rates and time delays are considered. The basis of the observer design methodology is a continuous-time nonlinear observer design. Subsequently, an inter-sample output predictor [33] is used to account for the slow-sampled measurements and to provide continuous estimates during the time period in between two consecutive measurements. At the same time, an estimate of the current output from the delayed measurement is obtained in the same spirit as the Smith predictor, by initializing the process model with the most recent delayed output and integrating it up to the present time. In the presence of sensor noise, a pre-filtering technique is used to cut out the noise to avoid the breakdown of the observer. The performance of the observer with inter-sample prediction and dead time compensation is evaluated by numerical simulation.

This paper is organized as follows: In Section 2, a brief review of the reduced-order observer and sampled-data observer design methods are presented. In particular, a block triangular observer form is derived from the serial subsystem structure (e.g., multiple continuous stirred-tank reactors (CSTRs) connected in series). In Section 3, the finishing stage of PET polycondensation, as well as its mathematical model is described. In Section 4, the performance of the state observer is evaluated in two different cases: (i) only continuous measurement is available; (ii) both continuous and slow-sampled measurements are available. Furthermore, sensor noise is considered, and the results show that there is a tradeoff between the convergence rate and noise sensitivity. Finally, in Section 5, conclusions are drawn from the results of the previous sections.

2. Nonlinear Observer Design Method

This section briefly outlines the main results on nonlinear observer design [30,31], block triangular observer design and sampled-data observer design [33]. All of the observer synthesis and simulations in later sections are realized on the basis of reduced-order observer. Therefore, a brief necessary review is presented below.

2.1. Reduced-Order Observer

In chemical processes, on-line measurements typically involve a part of the state vector. In contrast to the full-order observer, which estimates the entire state vector, the reduced-order observer estimates only the unmeasured states. In this sense, the reduced-order observer is free of redundancies and is computationally more efficient than the full-order observer.

Consider a multi-output autonomous system whose outputs are a part of the state vector:

$$\dot{x}_R = f_R(x_R, x_M)$$
$$\dot{x}_M = f_M(x_R, x_M) \tag{1}$$
$$y = x_M$$

where $x_R \in \mathbb{R}^{n-m}$ is the state vector that needs to be estimated, $x_M \in \mathbb{R}^m$ is the remaining state vector that is directly measured and $y \in \mathbb{R}^m$ is the measurement vector; $f_R : \mathbb{R}^n \to \mathbb{R}^{n-m}$ and $f_M : \mathbb{R}^n \to \mathbb{R}^m$ are real analytic functions with $f_R(0,0) = 0$, $f_M(0,0) = 0$. In the exact linearization method, the objective is to build an observer so that the resulting error dynamics is linear in curvilinear coordinates and with the pre-specified rate of decay of the error. A locally-analytic mapping $z = T(x_R, x_M)$ from $\mathbb{R}^n \to \mathbb{R}^{n-m}$ is sought that maps the system (1) to:

$$\dot{z} = Az + By \tag{2}$$

where A is a $(n - m) \times (n - m)$ matrix and B is a $(n - m) \times m$ matrix. The reduced-order observer in the original coordinates can be expressed as:

$$\dot{\hat{x}}_R = f_R(\hat{x}_R, y) + L(\hat{x}_R, y) \left(\frac{dy}{dt} - f_M(\hat{x}_R, y) \right) \tag{3}$$

This leads to the following selection of the state-dependent observer gain [31]:

$$L(\hat{x}_R, y) = - \left[\frac{\partial T}{\partial x_R}(\hat{x}_R, y) \right]^{-1} \frac{\partial T}{\partial x_M}(\hat{x}_R, y) \tag{4}$$

where $T(x)$ is a solution of the following system of PDEs:

$$\frac{\partial T}{\partial x_R} f_R(x) + \frac{\partial T}{\partial x_M} f_M(x) = AT + Bx_M \tag{5}$$

Under the above choice of observer gain, the error dynamics in transformed coordinates becomes linear and is governed by the arbitrarily-selected A matrix:

$$\frac{d}{dt}(T(x_R, y) - T(\hat{x}_R, y)) = A(T(x_R, y) - T(\hat{x}_R, y)) \tag{6}$$

Thus, the matrix A is a design parameter that directly adjusts the speed of convergence of the error.

Remark 1. *In order to implement the above nonlinear observer design methodology, an approximate solution needs to be calculated for the system of PDEs of Equation (5). As discussed in [30,31], it is possible to approximate $T(x_R, x_M)$ by using a truncated multivariable Taylor series around the origin. This requires each state expressed in deviation variable form. After expanding f_R, f_M and T in Taylor series up to a finite truncation order, the approximate solution can be obtained by equating the coefficient of each side of the PDEs. This calculation can be executed by using symbolic computation software (e.g., Maple) [31,34].*

2.2. Reduced-Order Observer in Lower Block Triangular Form

The serial CSTR reactor configuration is used in many types of chemical processes [35,36], leading to higher product yield and higher concentration. The serial CSTR reactor configuration usually possesses a special structure in lower block triangular (LBT) form. Additionally, this special structure can be utilized properly in state observer design to reduce the complexity of the state dependence of observer gains. Consider a system in LBT form containing three subsystems:

$$\dot{x}_{RI} = f_R^I(x_{RI}, x_{MI})$$
$$\dot{x}_{RII} = f_R^{II}(x_{RI}, x_{RII}, x_{MI}, x_{MII})$$
$$\dot{x}_{RIII} = f_R^{III}(x_{RI}, x_{RII}, x_{RIII}, x_{MI}, x_{MII}, x_{MIII})$$
$$y_I = x_{MI}$$
$$y_{II} = x_{MII}$$
$$y_{III} = x_{MIII}$$

$$\dot{x}_{MI} = f_M^I(x_{RI}, x_{MI})$$
$$\dot{x}_{MII} = f_M^{II}(x_{RI}, x_{RII}, x_{MI}, x_{MII})$$
$$\dot{x}_{MIII} = f_M^{III}(x_{RI}, x_{RII}, x_{RIII}, x_{MI}, x_{MII}, x_{MIII}) \quad (7)$$

where I, II, III denote each subsystem, respectively. The objective of observer design is to reconstruct the missing state variables x_{RI}, x_{RII} and x_{RIII}. Figure 1 depicts a general structure of the system in LBT form, with three subsystems.

It is intuitive to design sequential observers by taking advantage of the particular LBT structure. For example, the observer for Subsystem I is based on its unmeasured state dynamics and its own measurements y_I and is independent of the subsequent subsystems and their measurements. The observer for Subsystem II does not only use its own dynamics and measurements, but also depends on the measurements and state estimates from Subsystem I. Moreover, its state-dependent gain depends on the gain in the first observer, as well. Thus, each observer needs to utilize the information from all of the former stages, as well as its own dynamics and measurements. In this way, it significantly reduces the computational effort of calculating the state-dependent gain symbolically.

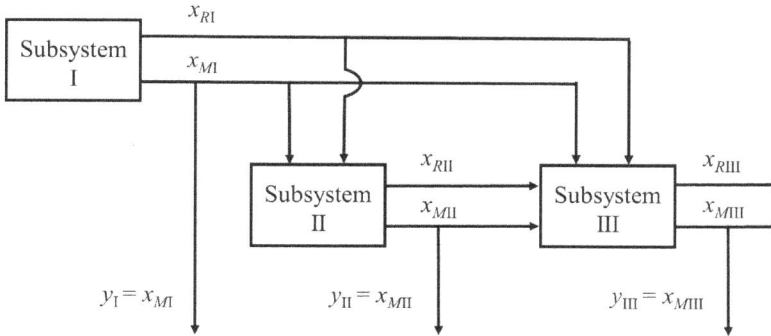

Figure 1. General structure of a system in lower block triangular (LBT) form with three subsystems.

After coordinate transformation, the observer in z-coordinates has linear dynamics:

$$
\begin{bmatrix} \dot{z}_I \\ \dot{z}_{II} \\ \dot{z}_{III} \end{bmatrix} = \begin{bmatrix} A_{11} & 0 & 0 \\ A_{21} & A_{22} & 0 \\ A_{31} & A_{32} & A_{33} \end{bmatrix} \begin{bmatrix} z_I \\ z_{II} \\ z_{III} \end{bmatrix} + \begin{bmatrix} B_{11} & 0 & 0 \\ B_{21} & B_{22} & 0 \\ B_{31} & B_{32} & B_{33} \end{bmatrix} \begin{bmatrix} y_I \\ y_{II} \\ y_{III} \end{bmatrix}
\tag{8}
$$

where both A and B matrix have a special LBT structure. Eigenvalues of the diagonal submatrices can be assigned arbitrarily. Since each subsystem's observer needs the estimates from the previous subsystems, it would make intuitive sense to tune the observer for Subsystem I faster than the one for Subsystem II, *etc.* Accordingly, the nonlinear reduced-order observer in original coordinates is of the form:

$$
\begin{bmatrix} \dot{\hat{x}}_{RI} \\ \dot{\hat{x}}_{RII} \\ \dot{\hat{x}}_{RIII} \end{bmatrix} = \begin{bmatrix} f_R^I(\hat{x}_{RI}, y_I) \\ f_R^{II}(\hat{x}_{RI}, \hat{x}_{RII}, y_I, y_{II}) \\ f_R^{III}(\hat{x}_{RI}, \hat{x}_{RII}, \hat{x}_{RIII}, y_I, y_{II}, y_{III}) \end{bmatrix} + \begin{bmatrix} L_{11} & 0 & 0 \\ L_{21} & L_{22} & 0 \\ L_{31} & L_{32} & L_{33} \end{bmatrix} \begin{bmatrix} \frac{dy_I}{dt} - f_M^I(\hat{x}_{RI}, y_I) \\ \frac{dy_{II}}{dt} - f_M^{II}(\hat{x}_{RI}, \hat{x}_{RII}, y_I, y_{II}) \\ \frac{dy_{III}}{dt} - f_M^{III}(\hat{x}_{RI}, \hat{x}_{RII}, \hat{x}_{RIII}, y_I, y_{II}, y_{III}) \end{bmatrix}
\tag{9}
$$

where the LBT state-dependent gain matrix $L(\hat{x}_R, y)$ can be designed according to:

$$
L_{11} = -\left[\frac{\partial T_1}{\partial x_{RI}}\right]^{-1} \frac{\partial T_1}{\partial x_{MI}}
$$

$$
L_{21} = -\left[\frac{\partial T_2}{\partial x_{RII}}\right]^{-1} \left[\frac{\partial T_2}{\partial x_{RI}} L_{11} + \frac{\partial T_2}{\partial x_{MI}}\right] \qquad L_{22} = -\left[\frac{\partial T_2}{\partial x_{RII}}\right]^{-1} \frac{\partial T_2}{\partial x_{MII}}
$$

$$
L_{31} = -\left[\frac{\partial T_3}{\partial x_{RIII}}\right]^{-1} \left[\frac{\partial T_3}{\partial x_{RI}} L_{11} + \frac{\partial T_3}{\partial x_{RII}} L_{21} + \frac{\partial T_3}{\partial x_{MI}}\right]
\tag{10}
$$

$$
L_{32} = -\left[\frac{\partial T_3}{\partial x_{RIII}}\right]^{-1} \left[\frac{\partial T_3}{\partial x_{RII}} L_{22} + \frac{\partial T_3}{\partial x_{MII}}\right] \qquad L_{33} = -\left[\frac{\partial T_3}{\partial x_{RIII}}\right]^{-1} \frac{\partial T_3}{\partial x_{MIII}}
$$

where $T(x) = \begin{bmatrix} T_1(x_{RI}, x_{MI}) \\ T_2(x_{RI}, x_{RII}, x_{MI}, x_{MII}) \\ T_3(x_{RI}, x_{RII}, x_{RIII}, x_{MI}, x_{MII}, x_{MIII}) \end{bmatrix}$ is a solution of the following system of PDEs:

$$
\frac{\partial T_1}{\partial x_{RI}} f_R^I + \frac{\partial T_1}{\partial x_{MI}} f_M^I = A_{11} T_1 + B_{11} x_{MI}
$$

$$
\frac{\partial T_2}{\partial x_{RI}} f_R^I + \frac{\partial T_2}{\partial x_{RII}} f_R^{II} + \frac{\partial T_2}{\partial x_{MI}} f_M^I + \frac{\partial T_2}{\partial x_{MII}} f_M^{II} = A_{21} T_1 + A_{22} T_2 + B_{21} x_{MI} + B_{22} x_{MII}
\tag{11}
$$

$$
\frac{\partial T_3}{\partial x_{RI}} f_R^I + \frac{\partial T_3}{\partial x_{RII}} f_R^{II} + \frac{\partial T_3}{\partial x_{RIII}} f_R^{III} + \frac{\partial T_3}{\partial x_{MI}} f_M^I + \frac{\partial T_3}{\partial x_{MII}} f_M^{II} + \frac{\partial T_3}{\partial x_{MIII}} f_M^{III}
$$

$$
= A_{31} T_1 + A_{32} T_2 + A_{33} T_3 + B_{31} x_{MI} + B_{32} x_{MII} + B_{33} x_{MIII}
$$

Under the above observer construction, the estimation error follows linear dynamics in z-coordinates, which is governed by the A matrix. It is selected to be Hurwitz to guarantee asymptotic stability.

2.3. Sampled-Data Observer

When sampling is performed at a slow rate, inter-sample behavior becomes important and needs to be accurately estimated by the observer. For this purpose, the process model could be used to predict the evolution of output during the time period in between two consecutive measurements. The predictor is able to continuously apply a correction on the most recent sampled measurement during the sampling interval.

The inter-sample output predictor can be combined with the reduced-order observer. The original system can be appropriately expressed in partitioned form as:

$$
\begin{aligned}
\dot{x}_R &= f_R(x_R, x_{Mc}, x_{Ms}) & y_c &= x_{Mc} \\
\dot{x}_{Mc} &= f_{Mc}(x_R, x_{Mc}, x_{Ms}) & y_s &= x_{Ms} \\
\dot{x}_{Ms} &= f_{Ms}(x_R, x_{Mc}, x_{Ms})
\end{aligned}
\tag{12}
$$

where $x_{Mc} \in \mathbb{R}^{m-1}$ is the state vector, which can be continuously measured, $x_{Ms} \in \mathbb{R}$ is the sampled state variable, and y_c and y_s are the corresponding outputs. Here, the output vector is split into two parts: $(m-1)$ continuous measurements and one sampled measurement.

It is possible to estimate the rate of change of the output $\dfrac{dy_s}{dt}$ by utilizing the dynamic model of slow-sampled state variable. This leads to the following inter-sample output predictor:

$$
\begin{aligned}
\frac{d\psi}{dt} &= f_{Ms}(\hat{x}_R, y_c, \psi), \quad t \in [t_k, t_{k+1}) \\
\psi(t_k) &= y_s(t_k)
\end{aligned}
\tag{13}
$$

with ψ representing the output prediction, and t_k, t_{k+1} denote two consecutive sampling instants. The predictor is initialized at the most recent measurement $y_s(t_k)$ and runs until the new measurement is obtained. When the continuous-time observer of Equation (3) is driven by the output predictor of Equation (13), this generates a sampled-data observer. Figure 2 depicts the construction of a continuous-time reduced-order observer with an inter-sample output predictor.

In earlier work [33], it was shown that, as long as the sampling period does not exceed a certain limit, the stability of the error dynamics and robustness with respect to measurement error for the continuous-time observer of Equation (3) implies the stability of the error dynamics and robustness with respect to measurement error for the sampled-data observer. In other words, the sampled-data implementation inherits the key properties of the continuous-time design, and in fact, these properties hold at all times, not just at the sampling instants.

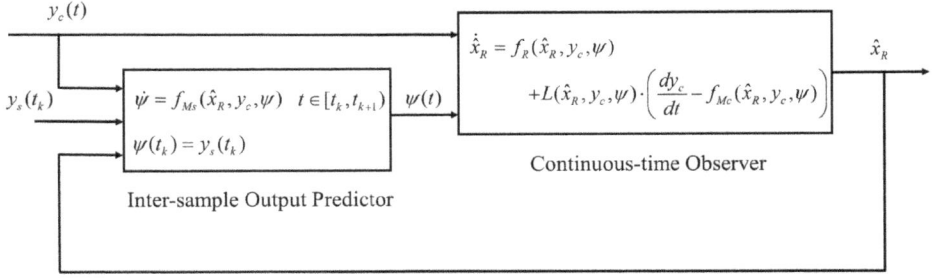

Figure 2. Structure of the reduced-order sampled-data observer.

3. A Series of Three Polycondensation Reactors

Modeling the finishing stage of PET synthesis is quite challenging due to the complexity of reaction kinetics, coupled with mass transfer effects. For the finishing stage, plug flow reactors (PFR) are commonly used because of their uniform residence time distribution, leading to a relatively narrow molecular weight distribution. In some continuous processes, a series of CSTRs are used [37]. The dynamics of plug flow polycondensation reactors can also be accurately modeled as multiple CSTRs in series [38].

For simplicity, a model of three CSTR in series, which is derived from Rafler's reaction-mass transfer model [32], will be used throughout this study. Figure 3 shows a three-CSTR in series configuration. In each reactor, the main polycondensation reaction and the thermal decomposition of ester groups are considered. Since the main reaction is reversible, EG as a byproduct has to be vaporized continuously by applying a vacuum to increase the yield of the product. The viscosity of the reaction mass also increases rapidly, which makes mass transfer a limiting factor. The dynamic process model has the following form:

Dynamics in CSTR I:

$$
\begin{aligned}
\frac{dx_1}{dt} &= \frac{1}{\tau_1}(x_{1,in} - x_1) - (\beta a)_1(x_1 - x_1^*) + \frac{1}{2}k_1(x_2{}^2 - 8x_1x_4) \\
\frac{dx_2}{dt} &= \frac{1}{\tau_1}(x_{2,in} - x_2) - k_1(x_2{}^2 - 8x_1x_4) \\
\frac{dx_3}{dt} &= \frac{1}{\tau_1}(x_{3,in} - x_3) + k_2x_4 \\
\frac{dx_4}{dt} &= \frac{1}{\tau_1}(x_{4,in} - x_4) + \frac{1}{2}k_1(x_2{}^2 - 8x_1x_4) - k_2x_4
\end{aligned}
\tag{14}
$$

Dynamics in CSTR II:

$$\frac{dx_5}{dt} = \frac{1}{\tau_2}(x_1 - x_5) - (\beta a)_2(x_5 - x_5{}^*) + \frac{1}{2}k_1(x_6{}^2 - 8x_5x_8)$$

$$\frac{dx_6}{dt} = \frac{1}{\tau_2}(x_2 - x_6) - k_1(x_6{}^2 - 8x_5x_8)$$

$$\frac{dx_7}{dt} = \frac{1}{\tau_2}(x_3 - x_7) + k_2x_8 \tag{15}$$

$$\frac{dx_8}{dt} = \frac{1}{\tau_2}(x_4 - x_8) + \frac{1}{2}k_1(x_6{}^2 - 8x_5x_8) - k_2x_8$$

Dynamics in CSTR III:

$$\frac{dx_9}{dt} = \frac{1}{\tau_3}(x_5 - x_9) - (\beta a)_3(x_9 - x_9{}^*) + \frac{1}{2}k_1(x_{10}{}^2 - 8x_9x_{12})$$

$$\frac{dx_{10}}{dt} = \frac{1}{\tau_3}(x_6 - x_{10}) - k_1(x_{10}{}^2 - 8x_9x_{12})$$

$$\frac{dx_{11}}{dt} = \frac{1}{\tau_3}(x_7 - x_{11}) + k_2x_{12} \tag{16}$$

$$\frac{dx_{12}}{dt} = \frac{1}{\tau_3}(x_8 - x_{12}) + \frac{1}{2}k_1(x_{10}{}^2 - 8x_9x_{12}) - k_2x_{12}$$

All three reactors are operated at constant temperature and pressure. There are four states in each reactor: the concentration of EG (x_1, x_5 and x_9), hydroxyl end groups (x_2, x_6 and x_{10}), carboxyl end groups (x_3, x_7 and x_{11}) and ester groups (x_4, x_8 and x_{12}). The concentration of EG on the melt surface is denoted by the superscript *.

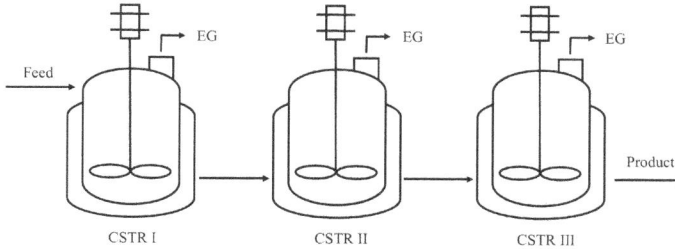

Figure 3. Schematic of three CSTRs in series in the polycondensation stage.

A two-film model is applied to describe mass transfer of volatiles in the finishing stage of melt polycondensation under high conversion. It is postulated that there is a concentration gradient of the volatile species throughout a liquid film near the gas-liquid interface. This is based on the existence of mass transfer resistance at the interface due to the high viscosity of the reaction mixture. Kim [39] verified the two-phase mass transfer model from experimental data in a polycondensation system

69

and showed that the mass transfer resistance model provided accurate prediction of molecular weight and product composition over the entire stages. The interfacial equilibrium concentration of EG is calculated by using the Flory-Huggins equation (see [39] for equations, [40,41] for physical property parameters). The system parameters used in the simulations are given in Table 1.

Table 1. System parameters [a,b].

Parameter	Description	Value
T	reactor temperature	553.15 K
P	reactor pressure	130 Pa
R	gas constant	1.987 cal/(mol·K)
$\tau_{1,2,3}$	residence time of each CSTR	60 min
k_1	rate constant of polycondensation reaction	$1.36 \times 10^6 \exp(-18,500/(RT))$ L/(mol·min)
k_2	rate constant of thermal decomposition	$7.20 \times 10^9 \exp(-37800/(RT))$ min^{-1}
$(\beta a)_1$	mass transfer parameter in CSTR I	2.70 min^{-1}
$(\beta a)_2$	mass transfer parameter in CSTR II	2.03 min^{-1}
$(\beta a)_3$	mass transfer parameter in CSTR III	1.35 min^{-1}

[a] k_1, k_2 are obtained from [41], and mass transfer parameter $(\beta a)_1$ is obtained from [42];
[b] mass transfer parameters in the last two reactors $(\beta a)_2$, $(\beta a)_3$ are assigned as follows: $(\beta a)_2 = 75\% \times (\beta a)_1$, $(\beta a)_3 = 50\% \times (\beta a)_1$.

In the reactor simulation, the following assumptions are made: (i) only EG exists in the vapor phase; (ii) mass transfer resistance on the gas side is negligible; (iii) the concentration of vinyl end groups in the feed is equal to the concentration of carboxyl end groups; (iv) the mass transfer parameter does not change over time in each reactor. The operating conditions of the reactors are given in Table 2, where [OH], [COOH] are for hydroxyl and carboxyl end groups and [Z] is the concentration of ester groups.

As pointed out in Section 1, the number of on-line measurements in polycondensation reactors is limited. Especially, measurements of various functional end groups are usually off-line, infrequent and delayed. In our study, two possible measurements are involved: one is continuous and the other is slowly sampled with dead time. The concentration of hydroxyl end groups can be obtained from a correlation using continuously-measured torque, temperature and stirrer speed, which needs to be calibrated for the specific reactor [28]. It can be considered as a continuous measurement without delay. The carboxyl concentration can be obtained by using acidimetric titration [44], which has a lower sampling rate and an approximately twenty-minute delay. DP is calculated from the state estimates using the formula:

$$DP = 1 + \frac{2[Z]}{[OH] + [COOH] + [E_v]} \qquad (17)$$

70

where $[E_v]$ denotes the concentration of vinyl end groups.

Table 2. Operating conditions and steady states [a,b].

Concentration	CSTR#	[EG]	[OH]	[COOH]	[Z]
Feed	CSTR I	6.5×10^{-3}	0.40	2.57×10^{-3}	11.2
	CSTR I	2.0×10^{-3}	0.40	2.57×10^{-3}	8.0
Initial Condition	CSTR II	1.0×10^{-3}	0.30	5.10×10^{-3}	8.0
	CSTR III	6.0×10^{-4}	0.24	6.31×10^{-3}	8.1
	CSTR I	5.645×10^{-4}	0.283	8.203×10^{-3}	11.25
Steady State	CSTR II	4.046×10^{-4}	0.226	1.385×10^{-2}	11.28
	CSTR III	3.470×10^{-4}	0.197	1.950×10^{-2}	11.28

[a] All of the concentrations are in units of mol/L; [b] feed condition is obtained from [43], which is the reactor outflow from the last stage (*i.e.*, pre-polymerization).

4. State Estimation via Reduced-Order Observer

Linear observability analysis was carried out in two different cases: (i) only hydroxyl end groups (x_2, x_6 and x_{10}) are continuously measured; (ii) in addition to hydroxyl end groups, carboxyl end groups (x_3, x_7 and x_{11}) are also measured by using on-line acidimetric titration. In Case (i), the conclusion is that the system is not observable, because carboxyl end groups are "downstream" relative to the hydroxyl end groups. It should be noticed that the interfacial concentration of EG does not depend on the state variables in the reactor. In Case (ii), the system of CSTRs is observable. The results of observability analysis suggest that the carboxyl measurement is necessary for accurate estimation of the states and, therefore, of DP, and it should be utilized in the observer despite its low sampling rate.

From a physical point of view, the system of CSTRs clearly possesses a serial structure: the outflow of the preceding reactor is the feed for the next reactor. Thus, it is straightforward to design sequential observers by taking advantage of the particular LBT system structure (as described in Section 2.2). The interconnection of these subsystems is shown in Figure 4, from which the unobservability in the absence of carboxyl measurements is clearly visible.

4.1. State Estimation with Continuous Measurement Exclusively

In Case (i), the output vector $y = \begin{bmatrix} x_2 & x_6 & x_{10} \end{bmatrix}^T$ represents the concentrations of hydroxyl end groups in the reactors, which are continuously measured. Even though the entire system is unobservable in the absence of carboxyl measurements, if only Subsystems Ia, IIa and IIIa are taken into account, the new system becomes observable. In other words, the concentrations of EG and ester groups can be estimated by using only hydroxyl measurement. For the specific system (*i.e.*, Ia, IIa and IIIa), we have implemented observer Equation (9) with state-dependent gain computed from Equation (10), where the mapping function $T(x)$ is a solution of the system of PDEs of Equation (11) with design parameters A and B. Two different choices of the A-matrix, with different sets of eigenvalues, are used in the simulations: "fast" $(-2.0, -1.8, -1.6, -1.4, -1.2, -1.0)$ and "slow" $(-0.2, -0.18, -0.16, -0.14, -0.12, -0.1)$. Truncation order N = 3 is used considering the balance between the accuracy of the approximate PDE solutions and computation time. The initial guess of the estimates is given in Table 3.

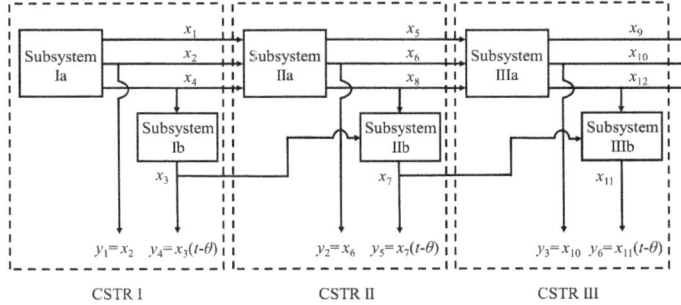

Figure 4. Subsystems representation of three CSTRs in series.

Table 3. Initial estimated values for the observer.

CSTR#	[EG] (mol/L)	[COOH] (mol/L)	[Z] (mol/L)
CSTR I	1.0×10^{-3}	7.57×10^{-3}	10.0
CSTR II	0	1.01×10^{-2}	10.0
CSTR III	1.6×10^{-3}	1.13×10^{-2}	10.1

As a result of being "downstream" states, carboxyl dynamics are detached from Subsystems Ia, IIa and IIIa. An open-loop observer is used to estimate the concentrations of carboxyl end groups, because their dynamics are open-loop stable. The open-loop observer equations are given as follows:

$$\frac{d\hat{x}_3}{dt} = \frac{1}{\tau_1}(x_{3,in} - \hat{x}_3) + k_2\hat{x}_4$$

$$\frac{d\hat{x}_7}{dt} = \frac{1}{\tau_2}(\hat{x}_3 - \hat{x}_7) + k_2\hat{x}_8 \qquad (18)$$

$$\frac{d\hat{x}_{11}}{dt} = \frac{1}{\tau_3}(\hat{x}_7 - \hat{x}_{11}) + k_2\hat{x}_{12}$$

with $\hat{x}_4$, $\hat{x}_8$ and $\hat{x}_{12}$ obtained from the observer equations driven by the continuous measurements y_1, y_2 and y_3.

Figure 5 shows the performance of the reduced-order observer with "fast" eigenvalues by comparing the actual and estimated states, as well as DP in the three CSTRs. As a result, the concentrations of EG and ester groups converge to the actual states very fast. Since the unobservable states (*i.e.*, concentrations of carboxyl end groups) are estimated from an open-loop observer, the speed of convergence depends on the dynamics itself, which is not adjustable. Therefore, it takes much longer to converge. This also explains the offset in the DP estimates in the beginning. However, this offset will be eliminated eventually as $\hat{x}_3$, $\hat{x}_7$ and $\hat{x}_{11}$ converge.

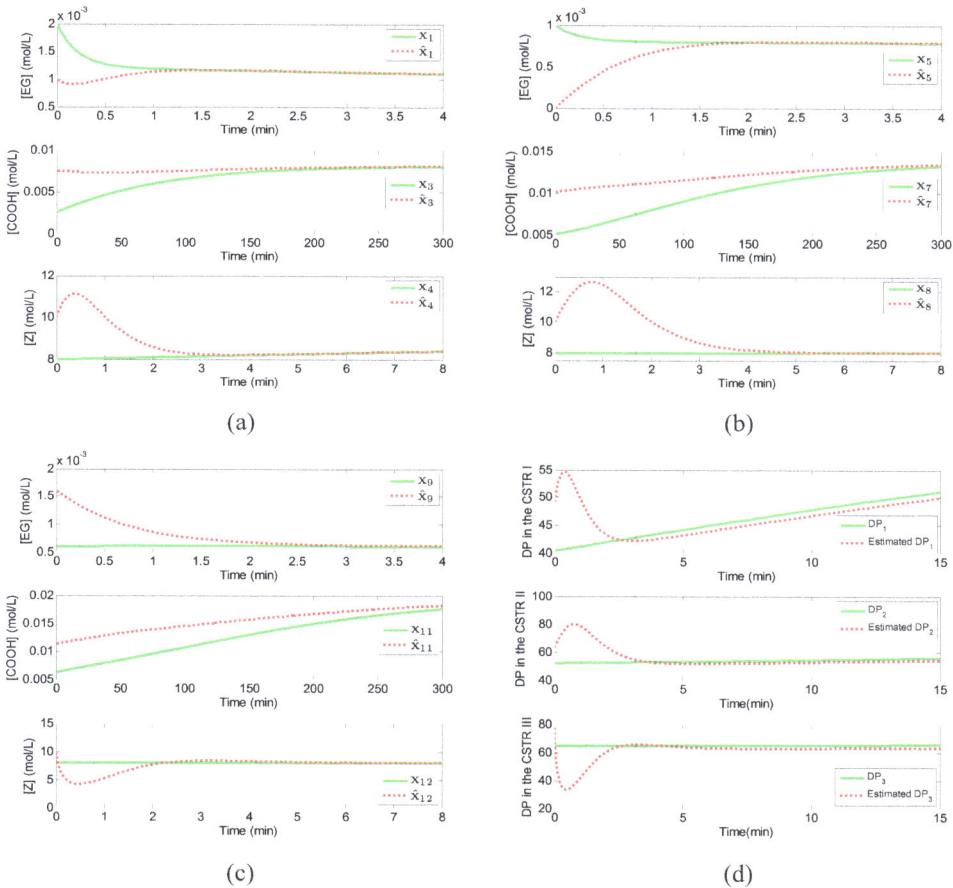

Figure 5. Performance of the reduced-order observer with "fast" eigenvalues: (**a**) actual and estimated states in CSTR I; (**b**) actual and estimated states in CSTR II; (**c**) actual and estimated states in CSTR III; (**d**) actual and estimated degree of polymerization in all three CSTRs.

4.2. State Estimation with Both Measurements

In Case (ii), both continuous and slow-sampled measurements are utilized in the observer design. Instead of using an open-loop observer, an inter-sample output predictor is used to estimate the evolution of the slow-sampled output during the sampling interval. Meanwhile, dead time compensation is carried out to account for the time delay between the present time and sensor dead time. For acidimetric titration, it is assumed that there is a ten-minute sampling interval, and the dead time of the sensor is twenty minutes. In this case study, it should be noticed that the output of the predictor does not need to feed into the reduced-order observer because carboxyl concentrations do not affect the other states and are not used in the estimation of concentrations of EG and ester groups. However, they will affect the estimation of DP. In this case, the dead time compensator is actually combined with the inter-sample output predictor. demonstrated as follows:

$$
\begin{aligned}
\frac{d\hat{y}_4}{dt} &= \frac{1}{\tau_1}(x_{3,in} - \hat{y}_4) + k_2\hat{x}_4, & t &\in [t_k - \theta, t_k + \eta) \\
\frac{d\hat{y}_5}{dt} &= \frac{1}{\tau_2}(\hat{y}_4 - \hat{y}_5) + k_2\hat{x}_8, & t &\in [t_k - \theta, t_k + \eta) \\
\frac{d\hat{y}_6}{dt} &= \frac{1}{\tau_3}(\hat{y}_5 - \hat{y}_6) + k_2\hat{x}_{12}, & t &\in [t_k - \theta, t_k + \eta) \\
\hat{y}_4 = y_4(t_k), \qquad \hat{y}_5 &= y_5(t_k), \qquad \hat{y}_6 = y_6(t_k)
\end{aligned}
\tag{19}
$$

where the state estimates $\hat{x}_4$, $\hat{x}_8$ and $\hat{x}_{12}$ are obtained from the continuous-time observer. y_4, y_5 and y_6 are the delayed outputs with dead time θ, while $\hat{y}_4$, $\hat{y}_5$ and $\hat{y}_6$ are the estimates at the present time, respectively. The three equations are initialized at the most recent measurement at t_k and run from $t_k - \theta$ to $t_k + \eta$, where η is the length of the sampling interval. It serves as a dead time compensator between $t_k - \theta$ and t_k and also serves as an inter-sample output predictor between t_k and $t_k + \eta$. In the first θ time units of each simulation, an open-loop observer is used for estimating carboxyl end groups, because there is no measurement information available.

In Figure 6, the convergence speed of EG and ester groups is slow because "slow" eigenvalues are chosen in this case. In the estimates of carboxyl concentrations, several steps are observed, because the slowly-sampled measurement corrects the predictor output when the most recent measurement becomes available each time. In addition, the observer together with the inter-sample predictor and the dead time compensator is able to estimate DP accurately in all three CSTRs.

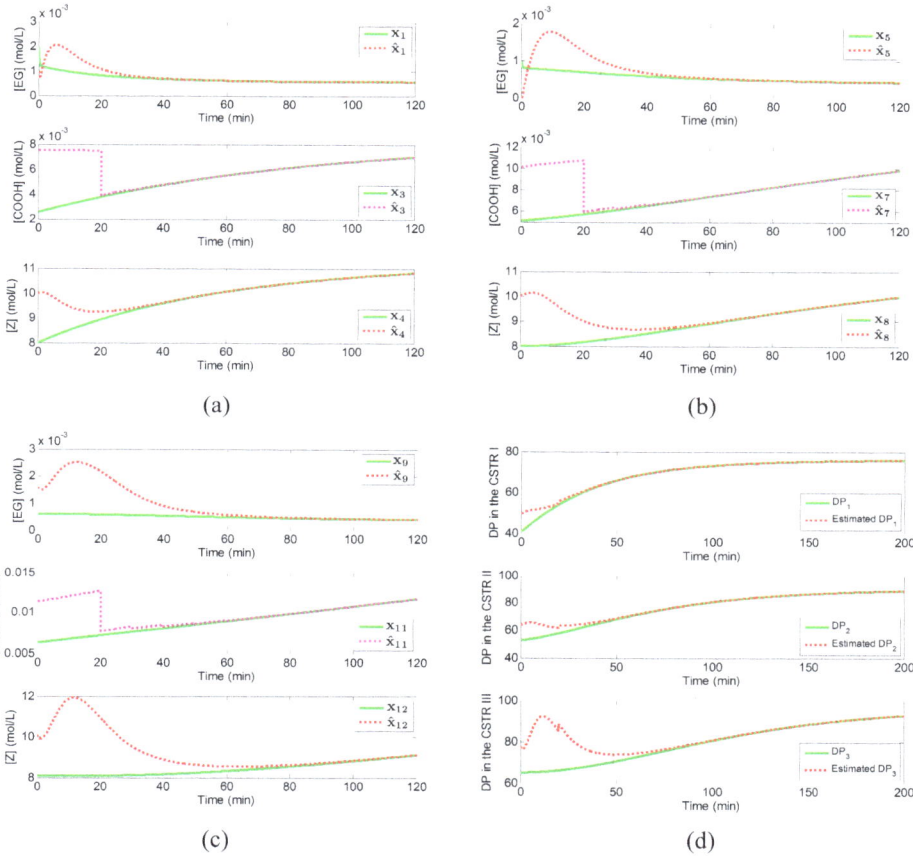

Figure 6. Performance of the reduced-order observer with "slow" eigenvalues when using both measurements: (**a**) actual and estimated states in CSTR I; (**b**) actual and estimated states in CSTR II; (**c**) actual and estimated states in CSTR III; (**d**) actual and estimated degree of polymerization in all three CSTRs.

4.3. Observer Performance under Sensor Noise

While the reduced-order observer is computationally more efficient by reconstructing only unmeasured state variables, it suffers from sensitivity to sensor noise. Therefore, the performance of the reduced-order observer needs to be tested under sensor noise. Pre-filtering of the measurement signal may be necessary to cut out the noise, which inevitably introduces some lag.

Figure 7 shows that the same level of white noise is added to all of the hydroxyl measurements with the standard deviation equal to 0.01. A first-order filter is used to cut out the high frequency noise. Different filter factors, 0.005, 0.005 and 0.007, are used respectively according to the filtering needed. As expected, we can see lags in

75

the filtered signal by comparing it to the actual value. White noise with standard deviation 3×10^{-4} is also considered for the on-line sampled titration measurements. Figure 7d shows both the estimated DP and the actual value when sensor noise is introduced. Fairly accurate estimation is achieved after about 70 min, even though the estimates deviate from the actual states quite significantly in the beginning. Relatively "slow" eigenvalues are used here because "fast" eigenvalues lead to a more aggressive response and may adversely affect observer performance.

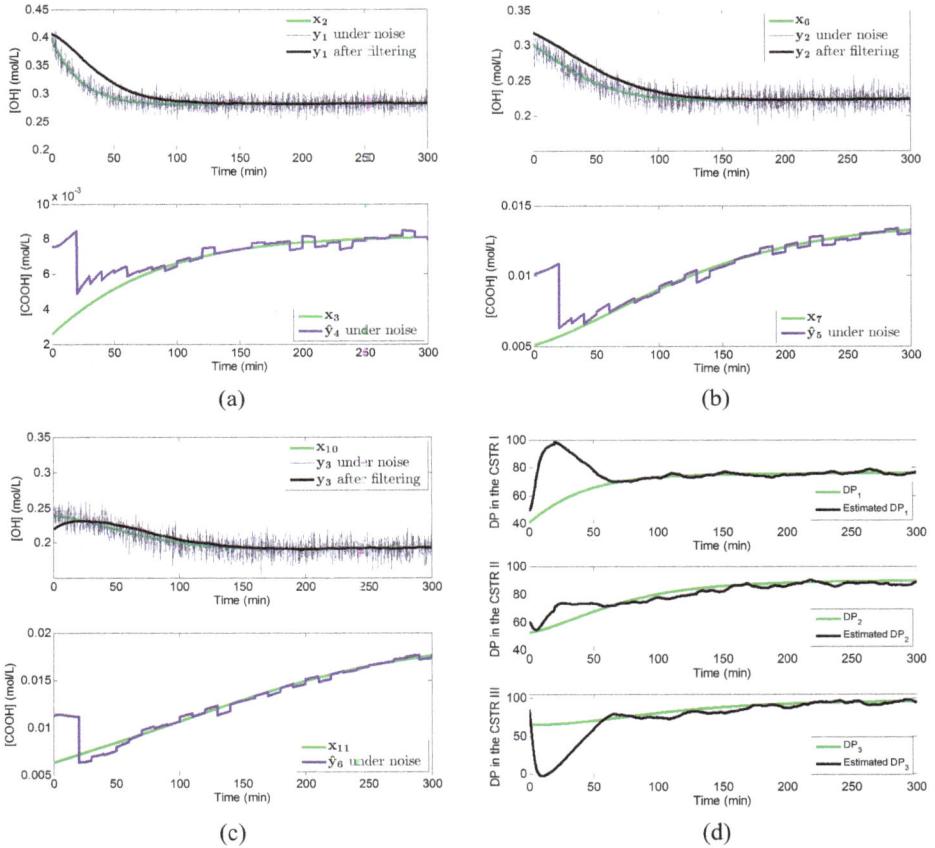

Figure 7. Measurement signals before (blue) and after (black) pre-filtering: **(a)** in CSTR I; **(b)** in CSTR II; **(c)** in CSTR III. Observer performance: **(d)** actual and estimated degree of polymerization in all three CSTRs.

5. Conclusions

This work presents an application of a nonlinear state observer for monitoring DP in a series of PET polycondensation reactors. By exploiting the special LBT

76

structure of the system, sequential observers are designed, and as a result, the complexity of the state dependence of observer gains is reduced. The unmeasurable states of EG and ester groups' concentrations are accurately estimated by using a reduced-order observer when only the continuous measurement is considered. The rate of convergence is adjustable by tuning the eigenvalues of design parameter A. When the slow-sampled measurement of carboxyl end groups is also available, an inter-sample output predictor is used to estimate the evolution of the sampled output during the sampling interval. Furthermore, dead time compensation is used to reduce the effect of delay in the output. Simulation results show that the degree of polymerization of PET is accurately estimated in all of the reactors when both continuous and sampled measurements are utilized. Even in the presence of sensor noise, the observer is still able to provide good estimates by applying pre-filtering.

Author Contributions: The conceptual framework was developed by Costas Kravaris. Chen Ling carried out simulations under the supervision of Costas Kravaris. Both authors were involved in the preparation of the manuscript.

Conflicts of Interest: The authors declare no conflict of interest.

References

1. Webbing Market Analysis By Product (Polyester, Nylon, Polypropylene, Carbon Fiber, Para Aramid Synthetic Fiber, UHMWPE), By Application (Automotive & Transport, Sporting Goods, Furniture, Military/Defense) And Segment Forecasts To 2020. Available online: http://www.grandviewresearch.com/industry-analysis/webbing-market-size (accessed on 10 January 2016).

2. Eldridge, J.E.; Ferry, J.D. Studies of the cross-linking process in gelatin gels. III. Dependence of melting point on concentration and molecular weight. *J. Phys. Chem.* **1954**, *58*, 992–995.

3. Daubeny, R.D.P.; Bunn, C.W. The crystal structure of polyethylene terephthalate. *Proc. R. Soc. Lond. A Math. Phys. Eng. Sci.* **1954**, *226*, 531–542.

4. McCormick, H.W.; Brower, F.M.; Kin, L. The effect of molecular weight distribution on the physical properties of polystyrene. *J. Polym. Sci.* **1959**, *39*, 87–100.

5. Torres, N.; Robin, J.J.; Boutevin, B. Study of thermal and mechanical properties of virgin and recycled poly(ethylene terephthalate) before and after injection molding. *Eur. Polym. J.* **2000**, *36*, 2075–2080.

6. Janssen, R.; Ruysschaert, H.; Vroom, R. The determination of the diethylene glycol incorporated in poly(ethylene terephthalate). *Makromol. Chem.* **1964**, *77*, 153–158.

7. Besnoin, J.M.; Choi, K.Y. Identification and characterization of reaction byproducts in the polymerization of polyethylene terephthalate. *J. Macromol. Sci. Rev. Macromol. Chem. Phys.* **1989**, *29*, 55–81.

8. Zimmerman, H.; Kim, N.T. Investigations on thermal and hydrolytic degradation of poly(ethylene terephthalate). *Polym. Eng. Sci.* **1980**, *20*, 680–683.

9. Adebekun, D.K.; Schork, F.J. Continuous solution polymerization reactor control. 2. Estimation and nonlinear reference control during methyl methacrylate polymerization. *Ind. Eng. Chem. Res.* **1989**, *28*, 1846–1861.

10. Jo, J.H.; Bankoff, S.G. Digital monitoring and estimation of polymerization reactors. *AIChE J.* **1976**, *22*, 361–369.

11. Kim, K.J.; Choi, K.Y. On-line estimation and control of a continuous stirred tank polymerization reactor. *J. Process Control* **1991**, *1*, 96–110.

12. Ellis, M.F.; Taylor, T.W.; Jensen, K.F. On-line molecular weight distribution estimation and control in batch polymerization. *AIChE J.* **1994**, *40*, 445–462.

13. Crowley, T.J.; Choi, K.Y. On-line monitoring and control of a batch polymerization reactor. *J. Process Control* **1996**, *6*, 119–127.

14. Mutha, R.K.; Cluett, W.R.; Penlidis, A. On-line nonlinear model-based estimation and control of a polymer reactor. *AIChE J.* **1997**, *43*, 3042–3058.

15. Dimitratos, J.; Georgakis, C.; El-Aasser, M.; Klein, A. Dynamic modeling and state estimation for an emulsion copolymerization reactor. *Comput. Chem. Eng.* **1989**, *13*, 21–33.

16. Kozub, D.J.; MacGregor, J.F. State estimation for semi-batch polymerization reactors. *Chem. Eng. Sci.* **1992**, *47*, 1047–1062.

17. Haseltine, E.L.; Rawlings, J.B. Critical evaluation of extended Kalman filtering and moving-horizon estimation. *Ind. Eng. Chem. Res.* **2005**, *44*, 2451–2460.

18. Dochain, D.; Pauss, A. On-line estimation of microbial specific growth-rates: An illustrative case study. *Can. J. Chem. Eng.* **1988**, *66*, 626–631.

19. Van Dootingh, M.; Viel, F.; Rakotopara, D.; Gauthier, J.P.; Hobbes, P. Nonlinear deterministic observer for state estimation: Application to a continuous free radical polymerization reactor. *Comput. Chem. Eng.* **1992**, *16*, 777–791.

20. Viel, F.; Busvelle, E.; Gauthier, J.P. Stability of polymerization reactors using I/O linearization and a high-gain observer. *Automatica* **1995**, *31*, 971–984.

21. Soroush, M. Nonlinear state-observer design with application to reactors. *Chem. Eng. Sci.* **1997**, *52*, 387–404.

22. Tatiraju, S.; Soroush, M. Nonlinear state estimation in a polymerization reactor. *Ind. Eng. Chem. Res.* **1997**, *36*, 2679–2690.

23. Sheibat-Othman, N.; Peycelor, D.; Othman, S.; Suau, J.M.; Fevotte, G. Nonlinear observers for parameter estimation in a solution polymerization process using infrared spectroscopy. *Chem. Eng. J.* **2008**, *140*, 529–538.

24. Tatiraju, S.; Soroush, M.; Ogunnaike, B.A. Multirate nonlinear state estimation with application to a polymerization reactor. *AIChE J.* **1999**, *45*, 769–780.

25. Astorga, C.M.; Othman, N.; Othman, S.; Hammouri, H.; McKenna, T.F. Nonlinear continuous–Discrete observers: Application to emulsion polymerization reactors. *Control Eng. Pract.* **2002**, *10*, 3–13.

26. Edouard, D.; Sheibat-Othman, N.; Hammouri, H. Observer design for particle size distribution in emulsion polymerization. *AIChE J.* **2005**, *51*, 3167–3185.

27. Choi, K.Y.; Khan, A.A. Optimal state estimation in the transesterification stage of a continuous polyethylene terephthalate condensation polymerization process. *Chem. Eng. Sci.* **1988**, *43*, 749–762.

28. Appelhaus, P.; Engell, S. Design and implementation of an extended observer for the polymerization of polyethylenterephthalate. *Chem. Eng. Sci.* **1996**, *51*, 1919–1926.

29. Yamada, T.; Imamura, Y.; Makimura, O.; Kamatani, H. A mathematical model for computer simulation of the direct continuous esterification process between terephthalic acid and ethylene glycol. Part II: Reaction rate constants. *Polym. Eng. Sci.* **1986**, *26*, 708–716.

30. Kazantzis, N.; Kravaris, C. Nonlinear observer design using Lyapunov's auxiliary theorem. *Syst. Control Lett.* **1998**, *34*, 241–247.

31. Kazantzis, N.; Kravaris, C.; Wright, R.A. Nonlinear observer design for process monitoring. *Ind. Eng. Chem. Res.* **2000**, *39*, 408–419.

32. Rafler, G.; Reinisch, G.; Bonatz, E.; Versaumer, H.; Gajewski, H.; Sparing, H.D.; Stein, K.; Mühlhaus, C. Kinetics of mass transfer in the melt polycondensation of poly(ethylene terephthalate). *J. Macromol. Sci.-Chem.* **1985**, *22*, 1413–1427.

33. Karafyllis, I.; Kravaris, C. From continuous-time design to sampled-data design of observers. *IEEE Trans. Autom. Control* **2009**, *54*, 2169–2174.

34. Kazantzis, N. Lie and Lyapunov Methods in the Analysis and Synthesis of Nonlinear Process Control Systems. Ph.D. Thesis, University of Michigan, Ann Arbor, MI, USA, 1997.

35. De Gooijer, C.D.; Bakker, W.A.; Beeftink, H.H.; Tramper, J. Bioreactors in series: An overview of design procedures and practical applications. *Enzyme Microb. Technol.* **1996**, *18*, 202–219.

36. Boe, K.; Angelidaki, I. Serial CSTR digester configuration for improving biogas production from manure. *Water Res.* **2009**, *43*, 166–172.

37. Thoenes, D. *Chemical Reactor Development: From Laboratory Synthesis To Industrial Production*, 1st ed.; Kluwer Academic Publishers: Dordrecht, The Netherlands, 1994.

38. Cao, L.; Yue, H. Modelling and control of molecular weight distribution for a polycondensation process. In Proceedings of the IEEE International Symposium on Intelligent Control, Taipei, Taiwan, 2–4 September 2004; pp. 137–142.

39. Kim, Y. Two phase mass transfer model for the semibatch melt polymerization process of polycarbonate. *Korean J. Chem. Eng.* **1998**, *15*, 671–677.

40. Daubert, T.E.; Danner, R.P. *Physical and Thermodynamic Properties of Pure Chemicals: Data Compilation*; Hemisphere Publishing Corporation: New York, NY, USA, 1989.

41. Ravindranath, K.; Mashelkar, R.A. Finishing stages of PET synthesis: A comprehensive model. *AIChE J.* **1984**, *30*, 415–422.

42. Kim, I.S.; Woo, B.G.; Choi, K.Y.; Kiang, C. Two-phase model for continuous final-stage melt polycondensation of poly(ethylene terephthalate). III. Modeling of multiple reactors with multiple reaction zones. *J. Appl. Polym. Sci.* **2003**, *90*, 1088–1095.

43. Bhaskar, V.; Gupta, S.K.; Ray, A.K. Modeling of an industrial wiped film poly(ethylene terephthalate) reactor. *Polym. React. Eng.* **2001**, *9*, 71–99.

44. Pohl, H.A. Determination of carboxyl end groups in polyester, polyethylene terephthalate. *Anal. Chem.* **1954**, *26*, 1614–1616.

Combining On-Line Characterization Tools with Modern Software Environments for Optimal Operation of Polymerization Processes

Navid Ghadipasha, Aryan Geraili, Jose A. Romagnoli, Carlos A. Castor, Jr., Michael F. Drenski and Wayne F. Reed

Abstract: This paper discusses the initial steps towards the formulation and implementation of a generic and flexible model centric framework for integrated simulation, estimation, optimization and feedback control of polymerization processes. For the first time it combines the powerful capabilities of the automatic continuous on-line monitoring of polymerization system (ACOMP), with a modern simulation, estimation and optimization software environment towards an integrated scheme for the optimal operation of polymeric processes. An initial validation of the framework was performed for modelling and optimization using literature data, illustrating the flexibility of the method to apply under different systems and conditions. Subsequently, off-line capabilities of the system were fully tested experimentally for model validations, parameter estimation and process optimization using ACOMP data. Experimental results are provided for free radical solution polymerization of methyl methacrylate.

Reprinted from *Processes*. Cite as: Ghadipasha, N.; Geraili, A.; Romagnoli, J.A.; Castor, C.A., Jr.; Drenski, M.F.; Reed, W.F. Combining On-Line Characterization Tools with Modern Software Environments for Optimal Operation of Polymerization Processes. *Processes* **2016**, *4*, 5.

1. Introduction

In the polymer industry, batch and semi-batch reactors are widely used for the production of different classes of polymers. There is considerable economic incentive to develop real-time optimal operating policies that will result in the production of polymers with desired molecular properties. In the case of polymerization processes, the molar mass distribution (MMD) of a polymer is one of the most important quality control variables since many of the polymer end-use properties are directly dependent on the MMD. Some examples include the mechanical properties such as stiffness, strength and viscoelasticity [1–3]. Optimal operation in polymerization usually involves computing and accurately maintaining the optimal policies that can lead to a product with desired MMD and final conversion, while minimizing the total operation time. However, direct feedback control of MMD is difficult to achieve due to several reasons. The primary issue is the lack of on-line promising method for monitoring polymer properties. Monitoring requires continuous measurement of the reacting solution to make data available for closed-loop control or state estimation of the system. The ability to monitor polymerization reactions as they occur is necessary to ensure fulfillment of the proposed recipe. Most of the control strategies in the literature for polymeric systems are based on open-loop methods which heavily depend on the accuracy of the mathematical model and will produce unsatisfactory results in case of disturbances in the system.

An extensive amount of work for feedback control of polymerization systems has been reported during last decades, most using an inferential technique to infer the polymer properties. Guyot *et al.* [4] used an on-line gas chromatographic analysis of the monomer mixture to control the copolymer composition in emulsion polymerization. Dimitratos *et al.* [5,6] developed a feedforward-feedback control strategy based on the use of an extended Kalman filter for state estimation. The Kalman filter method is based on a linear approximation of the nonlinear process [7] but has problems with stability and convergence [8–11]. For that reason many nonlinear methods have been developed. Hammori *et al.* [7] developed a nonlinear state observer that uses rate generation due to chemical reaction to obtain key parameters during free radical copolymerization. This technique is simpler to tune than Kalman filters. Kravaris *et al.* [12] utilized a general nonlinear feedforward-feedback control strategy based on temperature tracking to control copolymer composition. In order to control the nonlinear systems Model predictive control (MPC) [13–18] has been suggested. The main disadvantage of the MPC is that it cannot explicitly deal with plant model uncertainties.

Common manipulated variables in polymerization processes include coolant or heating medium flow rates, gas or liquid flow rates for pressure control, feed rate of monomer, solvent, or initiator, and agitator speed [19]. Crowley *et al.* [20] introduced a new method for controlling the weight chain length distribution of polymer in batch free radical polymerization processes by manipulating temperature. Although temperature has a considerable effect on MMD, when the reaction is extremely exothermic, which is the case in many polymerization systems, changing temperature is not an efficient way to control MMD. In this case, MMD is controlled by manipulating the ratio of monomer to initiator or adding a chain transfer agent. This will give the extra degree of freedom to minimize the operation time and also conversion of monomer. The combined control of temperature and reagent flow has been studied by Ellis *et al.* [21]. They proposed a MMD estimator based on an extended Kalman filter to provide current estimates of the entire MMD based on measurements of monomer conversion obtained by on-line densitometry and periodic time delayed measurements of MMD from on-line size-exclusion chromatograph. Simultaneous change of temperature and monomer addition to control MMD was investigated and it was demonstrated that the proposed feedback control strategy is effective in rejecting disturbances. In general, controlling molar mass is best achieved by manipulating the monomer, initiator and chain transfer agent concentration [5,22,23].

More recently a platform for automatic continuous online monitoring of polymerization (ACOMP) was proposed [24]. ACOMP allows for automatic, continuous and model independent monitoring of polymerization reactions. It uses continuous dilution of a small stream from the reactor, in conjunction with light scattering, viscosity, ultraviolet absorption and refractometric detectors. This provides monomer and co-monomer conversion, weight average molar mass, weight average intrinsic viscosity, average composition drift and distribution, and certain measures of polydispersity [24,25]. The ability to monitor polymerization reactions as they occur brings several benefits. First, following kinetics and other reaction characteristics allows fundamental understanding of reaction mechanisms that can help accelerate the development of new polymer chemistries and processes. Second, monitoring the effects of changing reaction conditions, such as temperature, reagent types, and concentration, provides a concise means for bench scale or small pilot reaction optimization.

This paper discusses the initial steps towards the formulation and implementation of a generic and flexible model-based framework for integrated simulation, estimation, optimization and feedback control of polymerization systems. The emphasis is on developing a comprehensive scheme which can be applied for the optimal operation of various polymeric systems and this was achieved by combining the powerful capabilities of the on-line monitoring system, ACOMP, with a modern simulation, estimation and optimization software environment. Using literature data, an initial validation of the framework for a number of cases of simulation and optimization is presented. This shows the flexibility of the method to work under different systems and conditions. Subsequently, the off-line capabilities of the framework were fully tested experimentally for model validations, parameter estimation as well as for process optimization. Starting with kinetic information from literature a new set of kinetic parameters were obtained for our case study as well as confidence intervals for the estimated parameters. This illustrates the importance of this strategy to follow kinetics and other reaction characteristics allowing fundamental understanding of reaction mechanisms that can help in the development of new polymer chemistries and processes. An offline model-based optimization analysis was conducted and experimentally validated to determine the optimal temperature profile in conjunction with the optimal monomer and initiator flow rate in order to reach a final target polymer while minimizing the batch time. A brief analysis on the controllability of the system under feedback conditions was also performed through a combination of simulation and experimental work.

2. Model Centric Framework

A shortcoming of using advanced operating and control strategies in polymerization processes is the unavailability of sensors able to determine online process status and provide corrective actions. The ACOMP platform will provide a leap towards integration of monitoring, control, and optimization tools in the control of complex industrial polymerization [26]. The proposed structure will forge initial links between ACOMP and advanced modelling and control principles and demonstrate unprecedented feedback control of polymerization reactions. The final goal is a self-contained intelligent system for advanced operation of polymerization processes where both ACOMP and the software environment exchange data seamless toward achieving target final products.

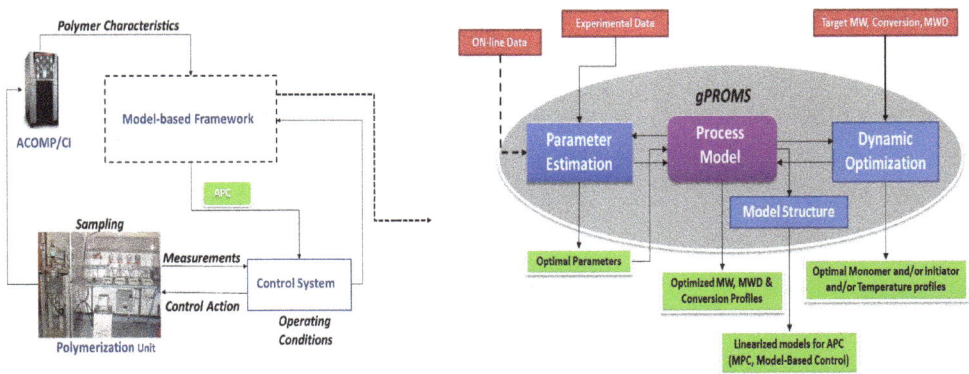

Figure 1. Schematic of the integrated simulation, estimation, optimization and feedback control of polymerization systems.

The conceptual representation of the aforementioned framework for integrated simulation, estimation, optimization and feedback control of systems is illustrated in Figure 1. The modelling work was carried out using gPROMS modelling language (V4.1.0, Process Systems Enterprise Inc London, United Kingdom, 2015), providing a complete environment for modelling/analysis of complex systems. Among gPROMS other advantages are: (a) modelling and solution power; (b) multiple activities using the same model; (c) integrated steady state and dynamic capabilities; (d) parameter estimation potentials; (e) sophisticated optimization tools; and (f) structural model information for advanced process control implementations. The parameter estimation entity makes use of the data gathered from the experimental runs. It has the ability to estimate an unlimited number of parameters, using data from multiple dynamic experiments and ability to specify different variance models among the variables as well as among the different experiments. The optimization

84

entity allows for the typical dynamic optimization problems arising from batch and/or semibatch operation to be formulated and implemented. One of the key issues is the connectivity of the software platform with the control system and ACOMP towards full integration. In this specific application gPROMS would need to be started from an external application and data will travel back and forth. An approach for this is to use the gSERVER API, transforming the application into the so-called gPROMS-based Application, or gBA or as an alternative via gO:RUN-xml. Both of these as well as open platform communication are currently explore in our experimental facilities.

2.1. Process Modelling

Modelling of polymerization processes is the other significant issue which plays a key role in the development of model-based control strategies. A number of well-known models have been introduced with much attention on the control of number and weight average molar mass [27,28]. Although weight average molar mass is usually the most important parameter in characterization of the polymer chain length distribution, there are cases in which controlling average properties may not be sufficient. For example, when a broad or bimodal distribution is desired [29]. A very attractive approach to characterize chain length distribution in batch free radical polymerization reactor has been presented by Crowley *et al.* [30,31]. This technique applies the method of finite molecular weight moments in conjunction with the kinetic rate equations to calculate the weight chain length distribution. As it was illustrated, it is feasible to control MMD in a batch polymerization process [20]. The modelling approach in this work incorporates the method of moment to obtain chain length distribution directly from the kinetic equations and includes the free volume theory to calculate the initiator efficiency, termination and propagation rate. As a first trial, a detailed mechanistic model for solution polymerization of methyl methacrylate (MMA) in batch and semi-batch reactors is developed and tested. This case study was selected since ample information regarding the kinetic mechanisms and data for MMA polymerization was available in the literature thus allowing faster prototyping. However, more complex systems will be considered in the future.

2.1.1. Reaction Mechanisms and Kinetic Equations

The reaction mechanisms adopted consists of three important steps: Initiation, propagation and termination. Chain transfer to monomer and solvent were also incorporated for better prediction of the molar mass. The detailed mechanism is as follow:

Initiation

$$I \xrightarrow{k_d} 2R \tag{1}$$

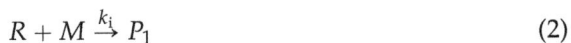

$$R + M \xrightarrow{k_i} P_1 \tag{2}$$

Propagation

$$P_n + M \xrightarrow{k_p} P_{n+1} \tag{3}$$

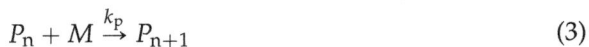

Chain transfer to monomer and solvent

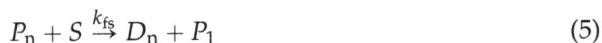

$$P_n + M \xrightarrow{k_{fm}} D_n + P_1 \tag{4}$$

$$P_n + S \xrightarrow{k_{fs}} D_n + P_1 \tag{5}$$

Termination

$$P_n + P_m \xrightarrow{k_{tc}} D_{n+m} \tag{6}$$

$$P_n + P_m \xrightarrow{k_{td}} D_n + D_m \tag{7}$$

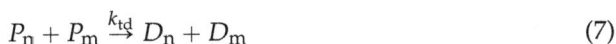

where I denotes the initiator, R is the primary initiator radical, M is the monomer, S is the solvent and P_j and D_j are the corresponding growing live polymer radical and dead polymer. Under standard assumptions such as well-mixed reactor, quasi steady state assumptions (for the radicals) and long chain hypothesis, the following set of kinetic and dynamic equations describe the system:

$$\frac{dN_m}{dt} = -\left(k_p + k_{fm}\right) P_0 N_m + F_m C_{mf} - F_{out} C_m \tag{8}$$

$$\frac{dN_i}{dt} = -k_d N_i + F_i C_{if} - F_{out} C_i \tag{9}$$

$$\frac{dN_s}{dt} = -k_{fs} N_s P_0 + F_i C_{sif} + F_m C_{smf} - F_{out} C_s \tag{10}$$

$$\frac{d(\lambda_0 V)}{dt} = (k_{fm} N_m + k_{td} P_0 V + k_{fs} N_s)\, \alpha P_0 + \frac{1}{2} k_{tc} P_0^2 V \tag{11}$$

$$\frac{d(\lambda_1 V)}{dt} = \left[(k_{fm} N_m + k_{td} P_0 V + k_{fs} N_s)\left(2\alpha - \alpha^2\right) + k_{tc} P_0 N\right] \frac{P_0}{(1-\alpha)} \tag{12}$$

$$\frac{d(\lambda_2 V)}{dt} = \left[(k_{fm} N_m + k_{td} P_0 V + k_{fs} N_s)(\alpha^3 - 3\alpha^2 + 4\alpha) + k_{tc} P_0 V (\alpha + 2) \frac{P_0}{(1-\alpha)}\right] \frac{P_0}{(1-\alpha)^2} \tag{13}$$

where $N_m = C_m V$, $N_i = C_i V$, $N_s = C_s V$

$$\alpha = \frac{k_p C_m}{k_p C_m + k_{fm} C_m + k_{fs} C_s + k_{tc} P_0 + k_{td} P_0} \tag{14}$$

$$P_0 = \sqrt{\frac{2 f C_i k_d}{k_{tc} + k_{td}}} \tag{15}$$

$$V = \left[1 - \frac{\lambda_1 w_m}{\rho_p}\right]^{-1} \left[\frac{N_m w_m}{\rho_m} + \frac{N_s w_s}{\rho_s} + \frac{N_i w_i}{\rho_i}\right] \qquad (16)$$

Here C_m, C_i and C_s represent the concentrations of monomer, initiator and solvent in the reactor, respectively. V illustrates the volume of the content of the reactor, F_m and F_i are the volumetric flow rate of monomer and initiator respectively which are fed into the reactor in the semi batch mode. F_{out} is the constant flow rate out of the reactor for the ACOMP extraction stream. C_{mf}, C_{if}, C_{sif} and C_{smf} are the concentration of monomer in the monomer feed stream, the initiator in the initiator feed stream and solvent in the initiator and monomer flow stream. P_0 is the total concentration of live polymer which is obtained from the quasi steady state assumption. λ_0, λ_1 and λ_2 are the corresponding moments for the dead polymers, and α is the probability of propagation. f is the initiator efficiency and ρ_m, ρ_i, ρ_s and ρ_p are the densities of the monomer, initiator, solvent and polymer which are temperature dependent. k_p, k_d, k_{fm}, k_{fs}, k_{tc} and k_{td} are the propagation, initiation, chain transfer to monomer, chain transfer to solvent, termination by combination and termination by disproportionation rate. Kinetic rate constants are all temperature dependent functions based on Arrhenius equation [28] and as we will see due to strong nonlinearity of the MMA system, the propagation and termination rate depend on conversion as well.

The conversion of the monomer is defined as the number of moles of monomer reacted in the tank divided by the total amount of monomer which has been loaded initially in the reactor and added by the semi-batch flow:

$$X = \frac{N_{m0} + \int_0^t F_m C_{mf} dt - C_m V - \int_0^t F_{out} C_m dt}{N_{m0} + \int_0^t F_m C_{mf} dt} \qquad (17)$$

Here, N_{m0} is the initial amount of monomer in the reactor. Number average and weight average molar mass of the polymers are calculated by considering only the moment of dead polymers and neglecting the live polymer concentration which is valid for low and medium conversions when the concentration of live polymer is negligible.

$$M_n = w_m \frac{\lambda_1}{\lambda_0}, \ M_w = w_m \frac{\lambda_2}{\lambda_1} \qquad (18)$$

2.1.2. Formalism for Gel, Glass and Cage Effects in MMA Polymerization

In free radical polymerization, the mobility of the radicals decreases along the reaction due to increase in the viscosity of the reactor as more polymers are produced. This phenomenon which is called "gel effect" causes a reduction in the termination rate constant k_t, and should be considered in the formulation of the model. At high conversion when even the motion of monomer is severely restricted the propagation rate k_p, is also decreased. This glassy state in which the solution is highly viscous sets a limiting conversion on the polymerization process. In this work the correlation by [28,32] is used for both gel and glass effect. This can be written as:

$$g_t = \begin{cases} 0.10575 \exp\left(17.15v_f - 0.01715\left(T - 273.15\right)\right), & v_f > v_{ftc} \\ 2.3 \times 10^{-6}\exp\left(75v_f\right), & v_f \leqslant v_{ftc} \end{cases} \tag{19}$$

$$g_p = \begin{cases} 1, & v_f > v_{fpc} \\ 7.1 \times 10^{-5}\exp\left(171.53v_f\right), & v_f \leqslant v_{fpc} \end{cases} \tag{20}$$

Here, v_{ftc} and v_{fpc} are the critical free volumes which are calculated as below:

$$v_{ftc} = 0.1856 - 2.965 \times 10^{-4}\left(T - 273.15\right) \tag{21}$$

$$v_{fpc} = 0.05 \tag{22}$$

v_f is the total free volume which is given by:

$$v_f = \phi_m v_{fm} + \phi_s v_{fs} + \phi_p v_{fp} \tag{23}$$

where

$$v_{fi} = 0.025 + \alpha_i\left(T - T_{gi}\right) \tag{24}$$

φ_i and T_{gi} are the volume fraction and the glass transition temperature of the polymer, solvent and monomer and α_i is a constant. The values of the parameter for this case are shown in Table 1 [28]. For butyl acetate the corresponding glass transition temperature was considered as an adjustable parameter which has to be determined by parameter estimation.

Table 1. Parameters for Methyl methacrylate gel and glass affect correlations.

Parameter	MMA	Poly Methyl Methacrylate	Butyl Acetate
α	0.001	0.00048	0.001
T_g(K)	167	387	#

The initiator efficiency f, appearing in Equation (15) describes the fraction of initiator free radicals which can successfully initiate the polymerization. Not all primary radicals can produce propagating chains. They execute many oscillations in "cages" before they diffuse apart and initiate a reaction. During the oscillations the radicals may also form an inactive species. Therefore, to account for the two competing phenomena, initiator efficiency is appended in the mathematical model.

Initiator efficiency factor also decreases as the viscosity of the reactor solution rises. The free volume theory is used to model this relationship:

$$f = f_0 \exp\left(-C\left(\frac{1}{V_f} - \frac{1}{V_{fcr}}\right)\right) \tag{25}$$

where f_0 is the initial initiator efficiency and C is a constant with the values of 0.53 and 0.006 for Azobisisobutyronitrile Fan *et al.* [33].

2.1.3. Molar Mass Distribution

In order to obtain complete representation of molecular weight distribution, a similar methodology proposed by Crowley *et al.* [30] based on finite weight fractions is applied with modifications for dealing with semi-batch operations. This method consists of dividing the entire polymer population into discrete intervals and calculating the weight fraction of polymer in each of the discrete intervals. By ignoring the concentration of live polymer, given that it is negligible compared with the dead polymer concentration, for each interval of (m,n) the polymer weight fraction is calculated with the following equations:

$$f(m,n) = \frac{\sum_{i=m}^{n} iD_iV}{\sum_{i=2}^{\infty} iD_iV} \text{ Or } f(m,n) = \frac{\sum_{i=m}^{n} iD_iV}{\lambda_1 V} \tag{26}$$

The dynamic of weight fraction can then be obtained as:

$$\frac{d f(m,n)}{dt} = \frac{1}{\lambda_1 V} \sum_{i=m}^{n} i\frac{d(D_iV)}{dt} - \frac{1}{\lambda_1 V} f(m,n)\frac{d(\lambda_1 V)}{dt} \tag{27}$$

where the right-hand side of Equation (27) represents the dynamic growth of dead polymers of length n which can be written according to the kinetic rate equation as below:

$$\frac{d(D_iV)}{dt} = \left[k_{fm}C_m + k_{fs}C_s + k_{td}C_p\right]P_nV \tag{28}$$

This can be shown by further simplification as follows:

$$\frac{d(D_iV)}{dt} = k_p C_m V P_i \frac{(1-\alpha)}{\alpha} \tag{29}$$

Substituting Equation (29) into (27) we get:

$$\frac{d f(m,n)}{dt} = \frac{1}{\lambda_1 V} k_p C_m V \frac{(1-\alpha)}{\alpha} \sum_{i=m}^{n} iP_i - \frac{1}{\lambda_1 V} f(m,n)\frac{d(\lambda_1 V)}{dt} \tag{30}$$

89

The term $\sum_{i=m}^{n} iP_i$ can be represented as:

$$\sum_{i=m}^{n} iP_i = \sum_{i=m}^{\infty} iP_i - \sum_{i=n+1}^{\infty} iP_i \tag{31}$$

Assuming, $P_n = \alpha P_{n-1}$ and $P_n = (1 - \alpha)\alpha^{n-1}P$:

$$\sum_{i=m}^{\infty} iP_i = \left[\frac{m(1-\alpha)+\alpha}{(1-\alpha)}\right]\alpha^{m-1}P - \left[\frac{(n+1)(1-\alpha)+\alpha}{(1-\alpha)}\right]\alpha^n P \tag{32}$$

And the final form of weight fraction for a semi-batch condition will be:

$$\frac{df(m,n)}{dt} = \frac{1}{\lambda_1}k_pC_m\left(\left[\frac{m(1-\alpha)+\alpha}{\alpha}\right]\alpha^{m-1} - \left[\frac{(n+1)(1-\alpha)+\alpha}{\alpha}\right]\alpha^n\right)P - \frac{1}{\lambda_1 V}f(m,n)\frac{d(\lambda_1 V)}{dt} \tag{33}$$

2.1.4. Energy Balances

One of the most complex features of the free radical polymerization is the exothermic nature of the reaction. Generated energy during polymerization should be removed by a coolant or dissipated to environment. Otherwise, the reactor can thermally run away. Even if run away does not occur, molar mass distribution can broaden. To model non-isothermal polymerization, energy balance should be applied to the reactant mixture in the reactor and oil in the bath. From the application of the energy conservation principle, the following equations show the energy balance for a perfectly mixed jacketed semi-batch reactor:

$$\frac{d(\rho_r C_{pr}VT_r)}{dt} = (-\Delta H)(k_p + k_{fm})N_m \times C_p - UA(T_r - T_j) + (F_m\rho_m C_{pm} + F_i\rho_s C_{ps})(T_f - T_r) \tag{34}$$

$$\frac{d(\rho_j C_{pj}V_jT_j)}{dt} = F_j\rho_j C_{pj}(T_{j,0} - T_j) + UA(T_r - T_j) \tag{35}$$

Here T_r and T_j denote the reactor and jacket temperature respectively. It was assumed that both reactor and jacket are perfectly mixed and have a constant temperature. ρ_r and C_{pr} are the average density and specific heat capacity of the reactor. C_{pm}, C_{ps} and C_{pj} are the specific heat capacity of monomer, solvent and coolant flow which consists of water and ethylene glycol. U is the overall heat transfer coefficient and A is the heat transfer area.

2.2. Parameter Estimation

Kinetic rate constants are significant parameters of a polymerization reaction which have to be determined accurately since even a slight change in them will result in considerable difference of the final polymer characteristics. The data regarding the kinetic rate constants may be obtained from literature or determined experimentally. However for some materials the properties are not available in the literature and it may not be possible to measure them through experiments due to lack of experimental facilities. Moreover, there are many criteria that affect the kinetic rate parameters such as reactor operating conditions, presence of inhibitors and purity of the materials which are different for various systems. So, proper values of the rate constants should be determined via a parameter estimation technique. This is an important prerequisite step in order to evaluate these variables and improve the model reliability for optimization and model-based control scheme development.

The parameter estimation is often formulated as an optimization problem in which the estimation attempts to determine the values for the unknown parameters in order to maximize the probability that the mathematical model will predict the values obtained from the experiments. Effective solution of parameter estimation is attainable if the following criteria are met [34]:

(a) The nonlinear system should be structurally identifiable which means that each set of parameter values will result in unique output trajectories.

(b) Parameters which have a weak effect on the estimated measured variables and the parameters which their effect on the measured output is linearly dependent should be detected and removed from the formulation of the estimation since their effect cannot be either accurately or individually quantified.

For the proposed system the parameters of the polymerization model that was introduced in section 2.1 can be represented as $z(t) = [X(t), M_w(t)]$ which are the outputs of the parameter estimation model, $u(t) = [T(t), F_m(t), F_i(t)]$ which are the time-varying inputs and θ the set of model parameters to be estimated which in this case are $[A_d, A_p, A_{td}, f_0, T_{gs}]$. A_d, A_p and A_{td} are the pre exponential factors of the decomposition rate, propagation rate and termination rate respectively. The selection of these parameters are justified as the most sensitivity in conversion and weight average molar mass data is with respect to the termination and propagation rate of a polymeric chain. Proper estimation of the initiator efficiency factor is also important since it controls the effective radical concentration. Since the transition temperature for butyl acetate is not available in the literature this parameter should also be estimated. In this work the parameter estimation scheme is based on maximum likelihood criterion. The gEST function in gPROMS is used as the software to estimate the set of parameters using the data gathered from the different experimental runs. Each experiment is characterized by a set of conditions under which it is performed, which are:

1. The overall duration.
2. The initial conditions which are the initial loading of initiator, solvent and monomer.

3. The variation of the control variables. For the batch experiment temperature is the only variable, while in semi batch both temperature and flow rate of monomer and/or initiator have to be considered.

4. The values of the time invariant parameters.

Assuming independent, normally distributed measurement errors, ϵ_{ijk} with zero means and standard deviations, σ_{ijk} this maximum likelihood goal can be captured through the following objective function:

$$\phi = \frac{N}{2}\ln(2\pi) + \frac{1}{2}\min_{\theta}\left\{\sum_{i=1}^{NE}\sum_{j=1}^{NV_i}\sum_{k=1}^{NM_{ij}}\left[\ln\left(\sigma_{ijk}^2\right) + \frac{\left(\tilde{z}_{ijk} - z_{ijk}\right)^2}{\sigma_{ijk}^2}\right]\right\} \tag{36}$$

where N describes the total number of measurements taken during all the experiments, θ is the set of model parameters to be estimated which may be subjected to a given lower and upper bound, NE, NV_i and NM_{ij} are respectively the total number of experiments performed, the number of variables measured in the ith experiment and the number of measurements of the jth variable in the ith experiment. σ_{ijk}^2 is the variance of the kth measurement of variable j in experiment i while $\tilde{z}_{ijk}$ is the kth measured value of variable j in experiment i and z_{ijk} is the kth model-predicted value of variable j in experiment i.

According to [35] the variable σ_{ijk}^2 depends on the error structure of the data which can be constant (homoscedastic) or depend on the magnitude of the predicted and measured variables (heteroscedastic). If σ_{ijk}^2 is fixed in the model the maximum likelihood problem is reduced into a least square criterion. If a purely heteroscedastic model applied the error has the following structure:

$$\sigma_{ijk}^2 = \omega_{ijk}^2\left(\tilde{z}_{ijk}\right)^\gamma \tag{37}$$

This means that as the magnitude of the measured variable increases the variance of $\tilde{z}_{ijk}$ also increases. The parameter ω_{ijk}^2 and γ are determined as part of the optimization during the estimation. In this work we assume the measurement error for both conversion and weight average molecular weight in all the experiments can be described by constant variance models since the errors for both conversion and weight average molecular weight is independent of their magnitude in the measurement. The given upper and lower bounds of the variance are based on the accuracy of the measurement plant and the function gEST specifies the ω_{ijk}^2 value along with X and M_w as part of the optimization.

2.3. Dynamic Optimization

The objective of the dynamic optimization is to find the optimal control profile for one or more control variables and control parameters of the system that drives the process to the desired final polymer property while minimizing the reaction time. The process control variables conform to their impact on the product quality and their capability for real time implementation. In this case temperature, monomer and initiator flow rate were selected as the control variables. Temperature plays a very important role in controlling the reaction kinetics which have a considerable impact on the molecular weight distribution while monomer and initiator flowrates are also a powerful means of controlling molar mass by affecting the concentration of the main feed to the reactor. The optimization of the model was performed using the gOPT function in gPROMS that applies the control vector parameterization (CVP) approach. Variation of the control variables in this case is considered as piecewise-constant, indicating the control variables remain constant at a certain value over a certain part of the time horizon before they jump discreetly to a different value over the next interval. The optimization algorithm determines the values of the controls over each interval, as well as the duration of the interval. Optimizer implements a "single-shooting" dynamic optimization algorithm consists of the following steps:

1. Duration of each control interval and the values during the interval are selected by the optimizer
2. Starting from the initial condition the dynamic system is solved in order to calculate the time-variation of the states of the system
3. Based on the solution, the values of the objective function and its sensitivity to the control variables and also the constraints are determined.
4. The optimizer revises the choices at the first step and the procedure is repeated until the convergence to the optimum condition is achieved.

For the proposed system the general optimal control problem is formulated as:

$$\min_{t_f, u(t), v} J(t_f) \tag{38}$$

Subjected to the process model and the following constraints:

$$x(t_0) - x_0 = 0 \tag{39}$$

$$t_f^{min} \leqslant t_f \leqslant t_f^{max} \tag{40}$$

$$u^{min} \leqslant u(t) \leqslant u^{max} \tag{41}$$

$$v^{min} \leqslant v(t_f) \leqslant v^{max} \tag{42}$$

where J for the general case is defined as:

$$J = w_1 \left(\frac{X_f}{X_t} - 1 \right)^2 + w_2 \left(\frac{M_{w,f}}{M_{w,t}} - 1 \right)^2 + w_3 \sum_{i=1}^{nc} \left(\frac{f_{i,f}}{f_{i,t}} - 1 \right)^2 + w_4 \left(\frac{t_f}{t_t} - 1 \right)^2 \quad (43)$$

Here x_0 is the initial condition of the system including the initial loading in the reactor and t_f stands for the time horizon while $u(t)$ indicates the control variables which are the temperature, monomer and initiator flow rates subjected to their lower and upper bounds. v_t represents the time variant parameters being the volume of the contents of the reactor. The formulation of the objective function consists of four terms. X_f, $M_{w,f}$ and $f_{i,f}$ are the values of the monomer conversion, molar mass and weight fraction of polymer within a chain length at the final time t_f respectively and X_t, $M_{w,t}$ and $f_{i,t}$ are their corresponding desired values. w_1–w_4 are the weighting factors, determining the significance of each term in the objective function. A schematic representation of the optimization problem for the polymerization problem is given in Figure 2.

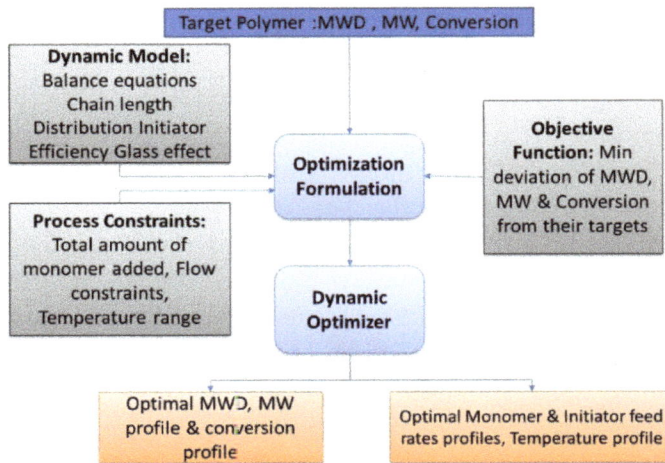

Figure 2. Schematic representation of the optimization problem in polymerization processes.

3. Experimental System

The methyl methacrylate monomer stabilized with 20 ppm of hydroquinone was supplied by Sigma-Aldrich. The Azo initiator, azobisizobutyronitrile (AIBN) was supplied by Sigma-Aldrich. Butyl acetate (BA) was used as solvent for the polymerization reactions and was supplied by Sigma-Aldrich. Nitrogen gas (N2) was supplied by Air Gas S.A. as an ultra-pure gas and used to keep the inert atmosphere inside the reactor. Acetone with minimum purity of 99.5% was supplied by Sigma-Aldrich and used to clean the reactor.

3.1. Experimental Apparatus—ACOMP System

Batch and Semi Bach solution polymerization reactions were conducted in the experimental setup shown in Figure 3. The reactor shown is a 540 mL cylindrical round-bottom glass reactor (Radleys, Shire Hill, Essex, United Kingdom), equipped with a double walled cooling jacket and reflux condenser. Agitation of the reactor (Custom Made at Tulane University PolyRMC, New Orleans, LA, USA) was done with a U-anchor style impeller driven by an overhead stirring motor (IKA Works, Inc., Wilmington, NC, USA). The reactor temperature was controlled with the aid of a thermostatic bath and a thermocouple (Huber USA, Inc., Cary, NC, USA) connected to a data acquisition system (Brookhaven Instruments Corp., Holtsville, NY, USA). The reflux condenser (Radleys, Shire Hill, Essex, United Kingdom), connected to a chilled thermostatic bath was used to prevent loss of reactants. To allow for continuous extraction of the reactor contents, the solution was circulated from the bottom of the reactor by a Zenith 9000 series gear pump (Colfax Corporation, Annapolis Junction, MD, USA) at 15 mL/min. From this circulation line a small stream was pulled by a Knauer high pressure isocratic pump (ICON Scientific Inc., North Potomac, MD, USA) and transferred to the first of two mixing chambers for dilution and conditioning of the polymer sample. This first stage of dilution is done under atmospheric pressure to allow for degassing of trapped air in the polymer stream. The now diluted and quenched polymer is again extracted by a Shimadzu LC-10ADvp isocratic pump (Shimadzu Scientific Instruments, Houston, TX, USA) and further diluted into a static mixing tee (Idex U-466S), (IDEX Health & Science LLC., Oak Harbor, WA, USA). Solvent inflow for both mixing chambers was provided by multiple Shimadzu LC-10ADvp isocratic pumps (Shimadzu Scientific Instruments, Houston, TX, USA). Now that the polymer stream has been thoroughly quenched, diluted and conditioned in the ACOMP "Front End" to a concentration level sufficient for intrinsic measurements, it is flowed through a series of detectors designated as the ACOMP "Detector Train". These detectors monitor the absolute properties of the polymer throughout the entire polymerization process.

The diluted sample stream now flows through a custom built single capillary viscometer (Custom Made at Tulane University PolyRMC, New Orleans, LA, USA), a Brookhaven Instrument BI-MwA multi-angle scattering photometer (Brookhaven Instruments Corp., Holtsville, NY, USA) (operating with a vertically polarized diode laser beam at 660 nm vacuum wavelength), a Shimadzu SPD-10AV dual wavelength UV/visible spectrophotometer (Shimadzu Scientific Instruments, Houston, TX, USA) and a Shimadzu RID-10A differential refractometer (Shimadzu Scientific Instruments, Houston, TX, USA). The signal from each instrument and the output of the reactor thermocouple were all digitized by A/D inputs on the BI-MwA input module, which contained 16 input channels working at 24 bit resolution. The UV wavelengths used was 269 nm. The viscometer used a capillary of length 10 cm and internal diameter of 0.02 in.

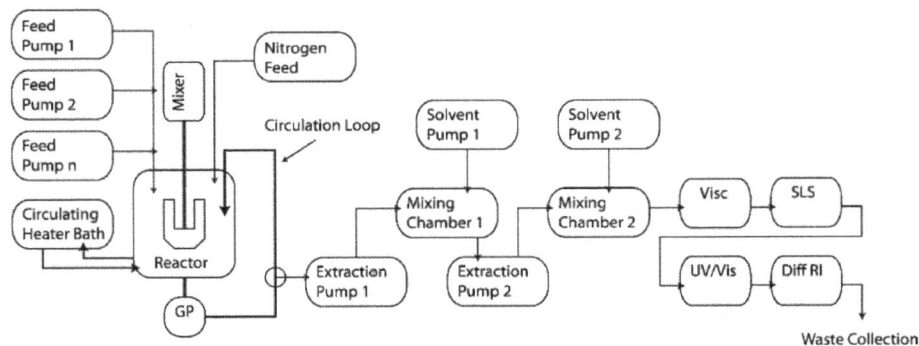

Figure 3. Automatic continuous on-line monitoring of polymerization setup for the monitoring of methyl methacrylate solution polymerization used in this work.

3.2. Experimental Procedure

The solvent in the reactor was purged with nitrogen at least 30 min prior to beginning the reaction. Before beginning the reaction, pure solvent (BA) was pumped at 2.0 mL/min through the entire detectors train (UV, RI, *etc.*) to obtain the baseline for each instrument. After stabilization of the detector baselines by pure solvent, the MMA (monomer) baseline was established by withdrawing from the reactor at 0.3 mL/min, and mixing with solvent at the mixing chamber. The flow rates of both the pure solvent from reservoir and that one from the reactor (containing the reaction mixture) used in this phase were 2.0 and 0.2 mL/min, respectively. Providing a dilution ratio of 1:10. In all of the experiments, the total flow rate was set at 2.0 mL/min and the diluted solution always reached the detector train at a temperature of 25 °C. After baseline stabilization for MMA the AIBN was added to initiate the reaction.

4. Results and Discussion

In this section, model predictions are compared with the literature data and experiments which have been performed using ACOMP system. The results are illustrated in terms of the conversion history and the evolution of molar mass and molar mass distribution. In the first case the proposed model is tested under a different system of initiator and solvent. The model is then verified against experimental data where the polymerization of methyl methacrylate using butyl acetate as solvent and AIBN as the initiator is considered.

4.1. Validation Using Literature Data

The experimental data provided by Crowley et al. [31] is used to investigates the applicability of the proposed framework. Batch free radical solution polymerization of methyl methacrylate has been carried out in a 4 L jacketed stirred tank reactor which was initially charged with 500 mL of MMA, 500 mL of ethyl acetate solvent, and 9 g of 2,2¢-azobis(2-methylbutanenitrile) initiator. The reactor was heated to 65 °C, regulated at that temperature to 27% monomer conversion, and then the temperature was decreased to 50 °C to intentionally broaden the molecular weight distribution. Results from the paper and simulation from the model are represented below for the conversion and molecular weight distribution at the end of the batch. As can be observed in Figure 4 they are in quite good agreement.

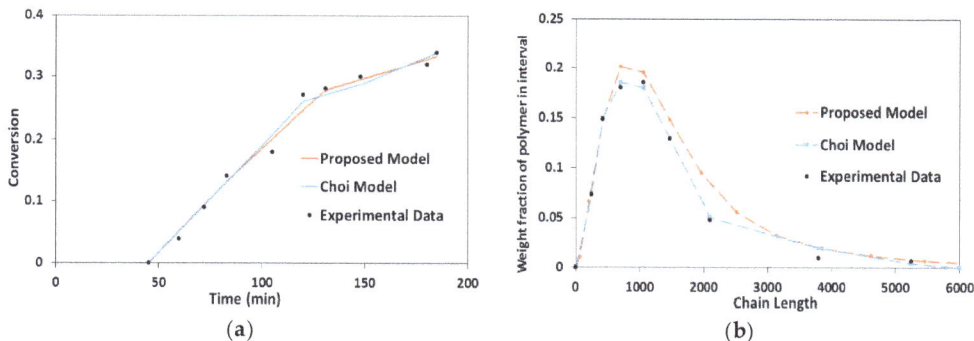

Figure 4. Comparison between model simulations and literature results. (**a**) Panel-Conversion profile. (**b**) Panel-Molar mass distribution at the end of the batch.

Data from Crowley et al. [31] was also utilized for the optimal batch operation. Based on the developed model it is possible to obtain the optimal trajectory of the temperature which lead to the desired molar mass distribution of polymers. The main problem in the batch optimization is the final time which is not specified. The proposed solution is to set all the kinetic equations with respect to conversion since its final value has already been specified. The solution of the optimization is a temperature profile with respect to conversion. The profile can then be implemented and the state of the system is monitored with respect to time. The objective now is to reach this final distribution by manipulating temperature before we reach the specified monomer conversion which is 0.7 in this case. There are two criteria that affect the results of the optimization. First, the number of intervals for the monomer conversion and second the initial guess for the temperature profile. Here the results obtained by manipulating the number of intervals are presented. Results for 1 and 5 intervals are provided in Figure 5, which illustrates the MMD for each case, as compared with the target. The results are in good agreement with those obtained in the original paper.

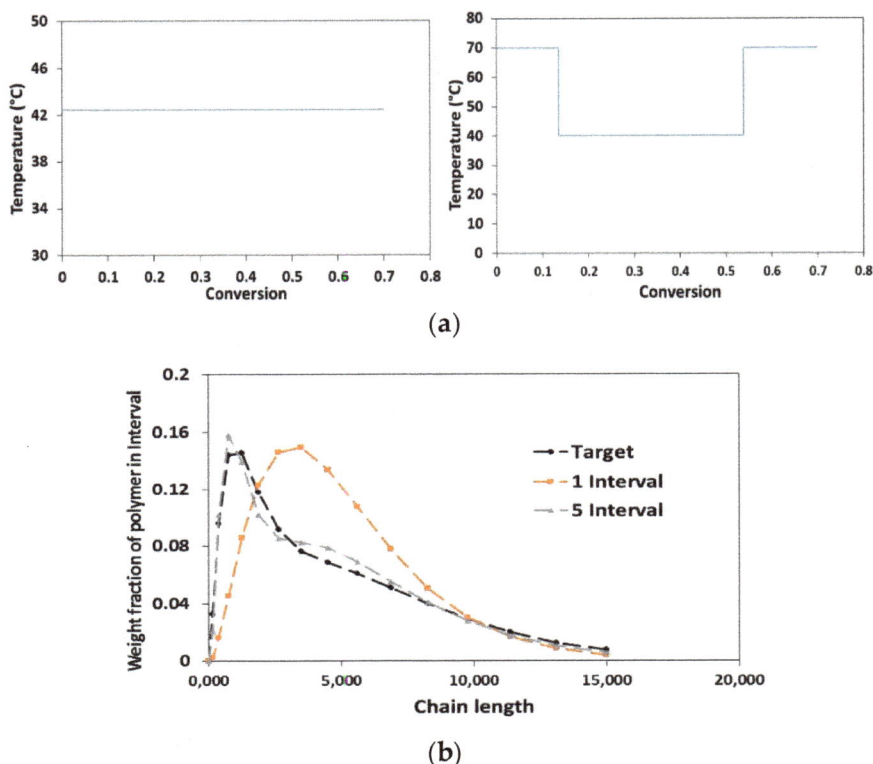

(a)

(b)

Figure 5. Optimization results: (a) Temperature profiles; (b) Molar mass distribution.

4.2. Experimental Validation for Batch and Semi-Batch Free Radical Polymerization of MMA Using Butyl Acetate as Solvent and AIBN Initiator

The experimental set-up discussed above was employed in this case to fully validate the proposed strategy for both batch and semibatch operation; Up to now previous works in the literature had been based on infrequent and incomplete sets of data and/or limited to batch operation. In this work model validations and parameter estimation have been carried out using ACOMP data. The capability of the model to properly describe the polymerization system has been investigated by doing a number of experiments in both semi-batch and batch mode using different trajectories for temperature and initiator and monomer flow rate. Conversion, weight average molar mass as well as molar mass distribution trajectories along the whole process were evaluated.

In an initial step, using a set of the experimental data, parameter estimation was performed for the main kinetic parameters. The optimal values of the estimated parameters as well as the uncertainty of the parameter represented as 95% confidence interval (CI) are shown in Table 2. Here, the contours correspond to a confidence level of 95%. In other words, there is a probability of 95% that the true values of the parameter pair fall within this ellipsoidal confidence region that is centered in the parameter estimates. In addition to confidence intervals the correlation matrix is also represented in Table 3 as a 5×5 lower triangular matrix (the upper triangular matrix is identical to the lower one). The most pronounced correlations between the parameters are shown in bold.

Table 2. Original and estimated value of the kinetic rate parameters for the free radical polymerization of MMA (first iteration).

Parameter	Description	Original Value	Estimated Value	Confidence Interval			95% t-value	Standard Deviation
				90%	95%	99%		
A_d	Decomposition (1/min)	1.58×10^{15}	1.37×10^{15}	1.25×10^{14}	1.49×10^{14}	1.96×10^{14}	9.19	7.60×10^{13}
A_p	Propagation (m^3/mol·min)	4.2×10^5	9×10^5	5.23×10^4	6.23×10^4	8.20×10^4	14.43	3.17×10^4
A_{td}	Termination [m^3/mol·min]	1.06×10^8	4.56×10^8	5.97×10^7	7.12×10^7	9.37×10^7	6.40	3.63×10^7
f_0	Initial Initiator Efficiency	0.58	0.57	0.048	0.057	0.076	9.84	0.029
T_s	Solvent Transition Temperature (K)	181	142.61	0.539	0.6431	0.84	221.7	0.327

Table 3. Correlation matrix for the estimated parameters of the system.

Estimated Parameters	A_d	A_p	A_t	f_0	T_s
A_d	1	-	-	-	-
A_p	**0.117**	1	-	-	-
A_t	**0.111**	**0.988**	1	-	-
f_0	**−0.988**	**0.038**	**0.043**	1	-
T_s	**0.104**	**0.179**	**0.233**	**−0.070**	1

Most of the estimated parameters obtained have narrow confidence intervals indicating that the number of measurements performed for the parameter estimation were sufficient. The normalized covariance matrix shows that although a few parameters are quite correlated, most parameters estimated in the optimization are only weakly correlated and therefore are suitable for being estimated simultaneously. Larger correlation coefficients are found between the propagation rate and the termination rate as well as initial initiator efficiency and decomposition rate as shown in Figure 6. Likely, any change in one of these parameters could be compensated by a change in the other ones. For example, the coefficient between A_p and A_{td} is 0.94 indicating a strong correlation between them and making it difficult to find a unique estimate for these parameters. Unique parameter estimate means that the parameters have an acceptably low correlation to any of the other parameters and a low confidence interval. Thus, in spite of the large covariance mentioned above, a consistent estimation is possible because of the true value of the estimated parameters are located within a very small confidence bands reducing their uncertainty. However, the confidence ellipsoids are large including in most cases negative numbers. Therefore, a second iteration was performed by eliminating two of the correlated parameters A_d and A_{td} and fixing their values to the estimated ones in the first iteration. The optimal values of the estimated parameters as well as the uncertainty of the parameter represented as 95% confidence interval (CI) are shown in Table 4.

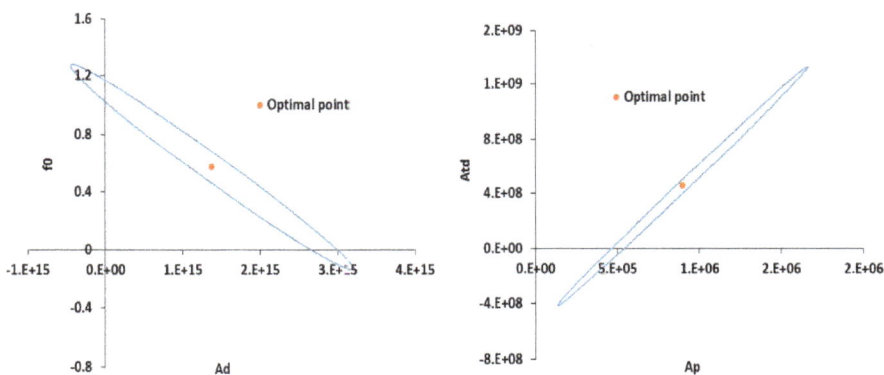

Figure 6. Confidence ellipsoids for A_d–f_0 and A_p–A_t.

For purposes of illustration the confidence regions for the parameter pairs estimated are shown in Figure 7. None of the confidence ellipsoids cross either the x or y axes. Thus, no parameter pair has a parameter value equal to zero. The confidence ellipsoids also show small negative correlation between propagation constant and corresponding glass transition temperature for the solvent.

100

Table 4. Original and estimated value of the kinetic rate parameters (second iteration) for the free radical polymerization of MMA.

Parameter	Description	Original Value	Estimated Value	Confidence Interval			95% t-value	Standard Deviation
				90%	95%	99%		
A_p	Propagation Rate $(m^3/mol \cdot min)$	3×10^5	8.5×10^5	2547	3035	3993	280.1	1546
f_0	Initial Initiator Efficiency	0.58	0.56	0.001166	0.0013	0.00182	403.2	0.00073
T_s	Solvent Transition Temperature (K)	142	149.94	0.3906	0.465	0.6123	322.2	0.237

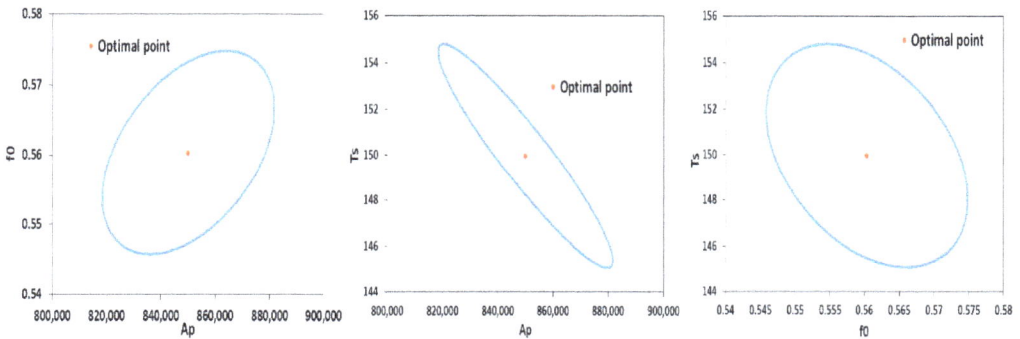

Figure 7. Confidence ellipsoids for final estimated parameters.

Figure 8 shows the conversion and molar mass profiles for two batch conditions with constant and varying temperature. As can be observed, the parameter estimation significantly improves the results. The simulation results after the parameter estimation have an excellent agreement with the experimental data for both conditions, indicating the adjusted model has good predictive capabilities for the proposed system.

Next, the adjusted model is embedded into the optimization environment to investigate alternative optimal operational policies for final target products. Two objective functions were formulated. One is to determine the optimal trajectories of the control input values that minimize the reaction time while the product qualities reach the specification at the end of the process and the other will only consider the product properties at the end of the process given enough time to the reaction. The optimizer iteratively computes the sequence of reactor temperature, monomer and initiator flow rates which will yield the best match between the final conversion, molar mass and molar mass distribution and their corresponding target values. Figure 9 presents a snapshot of the three selected iterations. The graph on the left-hand side shows the conversion profile and the right diagram represents the calculated objective function values. Since the solution for early iterations are not optimal the final objective function is high and there is a large discrepancy in the final conversion. Simulation results of optimal profiles at final iteration are shown in Figure 10 for the three control variables and as can be observed there is an excellent agreement between the final values and the target suggesting the advantage of using reagent and monomer flow with temperature to control not only the conversion and weight average molecular weight but also the complete distribution. It should be noted that the optimizer, CVP-S (Control Vector Parameterization- Single Shooting), used by gPROMS may find the local minima rather than the global one. So, the solution is sensitive to the initial point and it is necessary to properly initialize the optimization problem in order to capture the global minima in case of local minima existence.

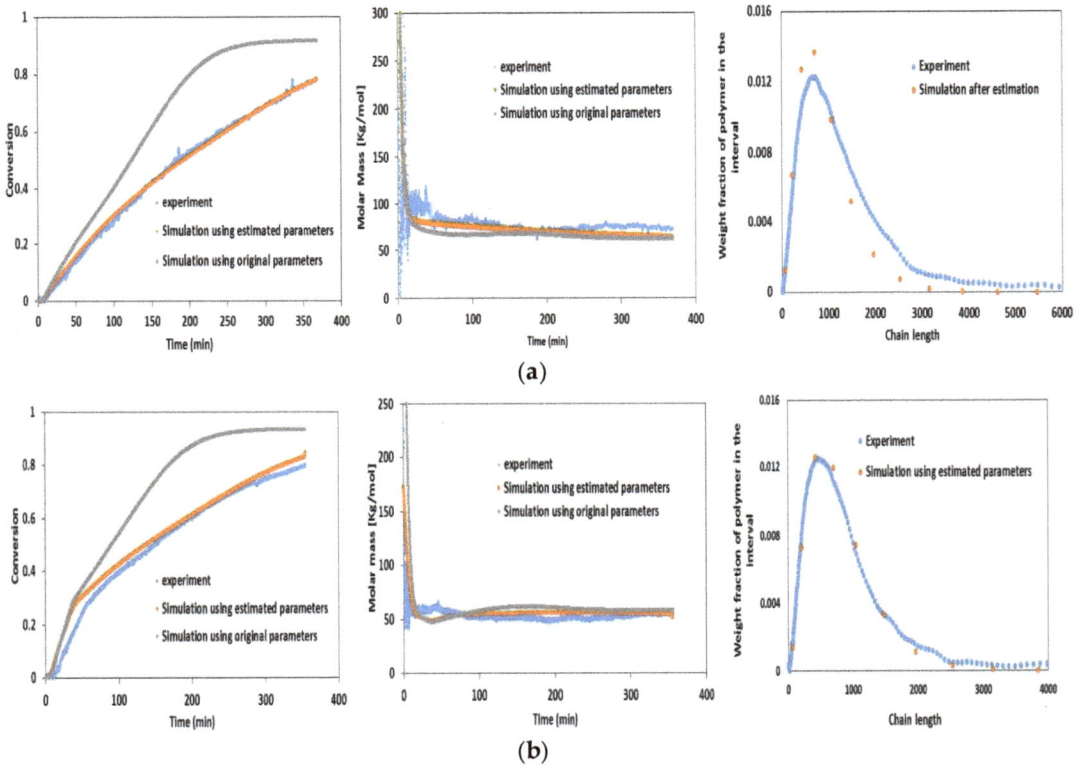

Figure 8. Comparison between experimental data and simulation with original and estimated parameters: (**a**) Isothermal experiment; (**b**) Non-isothermal experiment.

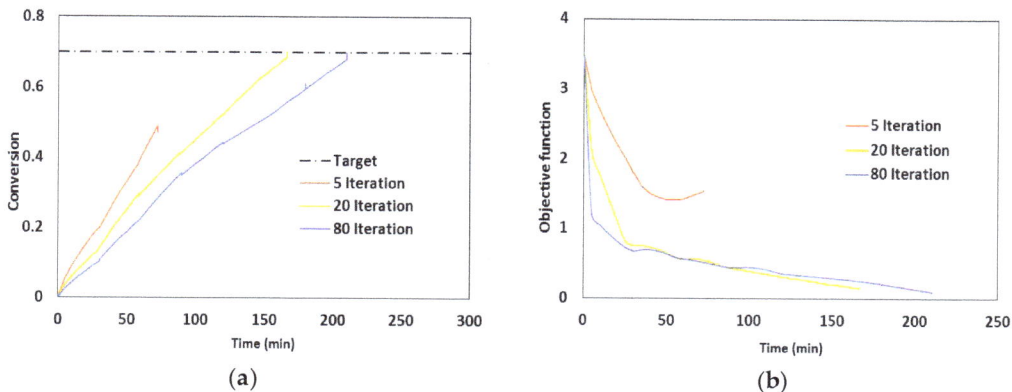

Figure 9. Conversion (**a**) and objective function (**b**) profiles at three different iterations.

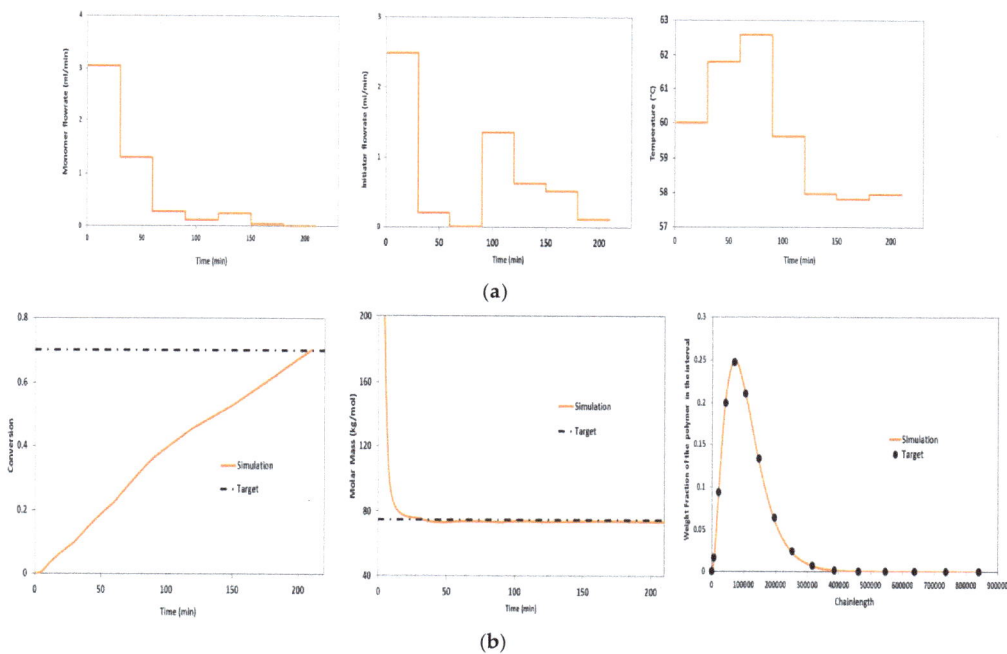

Figure 10. Simulation results of the optimal trajectories considering time in the objective function: (**a**) Input (manipulated) variables; (**b**) Controlled variables (targets).

To demonstrate the feasibility of the proposed optimization strategy the second optimization algorithm for the objective function has been validated also experimentally using data from ACOMP. Table 5 provides the values of the different parameters and the constraints used in this experiment. Temperature and flow constraints are used based on the capacity of the pump and jacket while the minimum volume constraint is the minimum necessary volume for the sensors to have a good estimation of the reactor condition. Figure 11 shows the resulting optimal trajectories of the input variables (temperature, monomer and initiator flows) with the validation results in terms of the model predictions (targets) and the experimental data when the obtained inputs trajectories are applied into the experimental systems. The complete distribution has been shown at the final time of the experiment. The maximum deviations from the model simulation is in the order of 0.03 for conversion and 5 kg/mol for the molar mass. The final distribution is also in a very good agreement with the model predictions showing the feasibility of the proposed scheme.

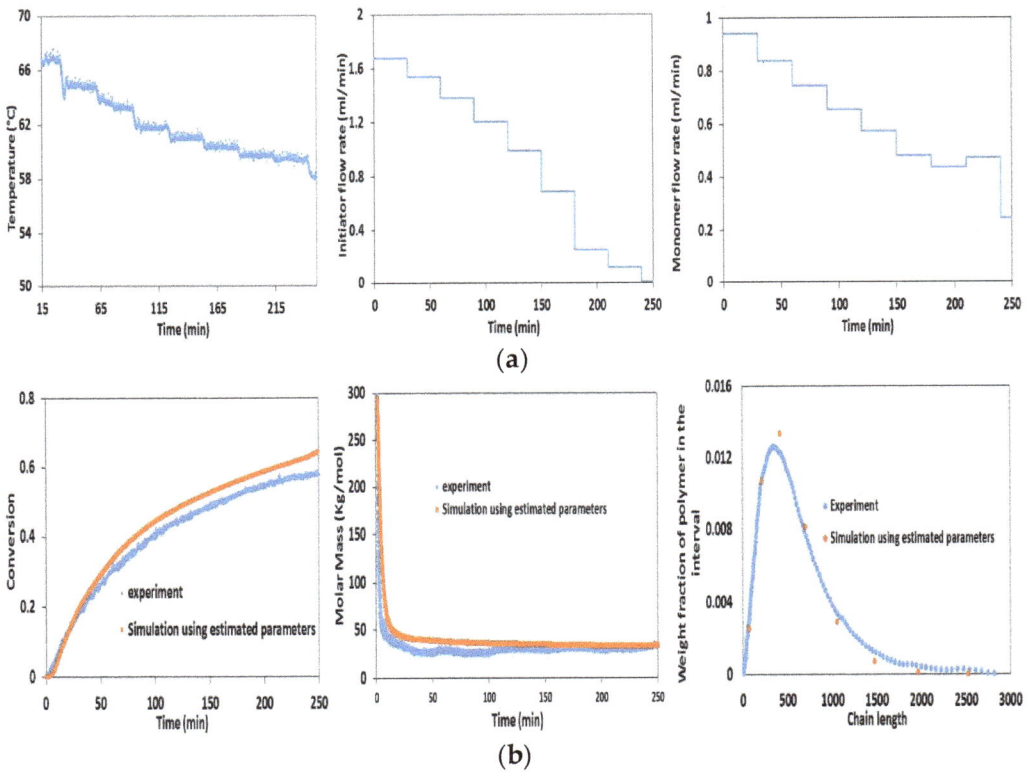

Figure 11. Validation of optimal runs: **(a)** Input (manipulated) variables; **(b)** Controlled variables (targets).

Table 5. Values of optimization constraints parameters.

Variable	Value	Unit
N_m	0.5	mol
N_s	0.5	mol
N_i	0.01	mol
F_{max}	5	mL/min
F_{min}	0	mL/min
T_{max}	70	°C
T_{min}	50	°C
V_{max}	500	mL
V_{min}	100	mL

To analyze the controllability of the system under feedback control and as a preliminary step towards the on-line feedback control implementation, a number of simulations combined with experimental information were performed. In this study, the experimental data from the optimal semi-batch operation for the corresponding controlled variable was used as a set-point in a simulated feedback scheme. In this way the experimental optimal profiles for conversion (in the single loop case) and experimental optimal profiles of both conversion and molar mass flow (in the multi-loop case) were provided to the simulator/controller to adjust the temperature and both temperature and initiator flow to achieve the targets. The adjusted input trajectories are then compared to the experimental ones showing the necessary adjustments needed for the control system to care for the model uncertainties. The objective in this part is to propose a straightforward modification to the open-loop optimal recipe described above that will allow the process to meet the optimal condition under non-ideal process circumstances.

The block diagram of the proposed structure is shown in Figure 12. The three possible manipulated variables in this case are the temperature, monomer and initiator flow rate. In the single loop approach, two of the possible manipulated variables still follow the same optimal recipe while the other variable changes based on a PI-like control algorithm in order to follow exactly the set point trajectory. In this case conversion has been selected as the control variable and its controllability with respect to the reactor temperature is investigated. The temperature adjusted using a cascade control system which has been represented schematically in Figure 13. The approach followed here is to start the operation in an open-loop fashion and switch to feedback control during the batch operation. The switching time is around 20 min for the single loop and 60 min for the multi loop controller. Figure 14 shows the performance of the closed loop system. As can be seen good control of the conversion through the batch operation was attained by slight modifications of the open-loop optimal trajectories as predicted from the optimization step. However the molar mass is still out of specifications with respect to the desired final value.

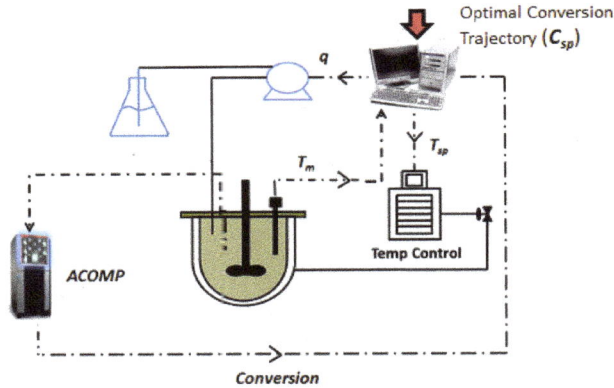

Figure 12. Schematic of apparatus and sensors for the proposed control strategy.

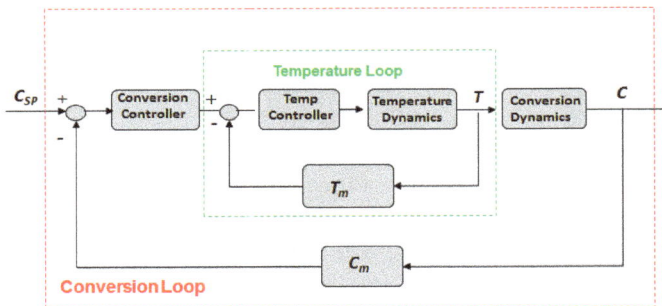

Figure 13. Schematic of cascade control for conversion.

In the multi-loop approach the conversion was again controlled using the cascade configuration as described above but an additional loop was added by adjusting the initiator flow to control the molar mass. Figure 15 illustrates how both conversion and molar mass trajectories follow exactly the desired values as indicated in the plot. One should notice however that small corrections in the molecular weight requires a large control action (initiator flow) indicating low sensitivity of the manipulated variable with respect to the control variable. Also, the interaction between the two loops can be appreciated since the temperature and conversion profiles are slightly modified with respect to the single loop case. Specially during the first action of the initiator controller.

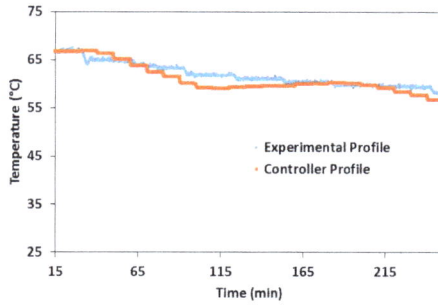

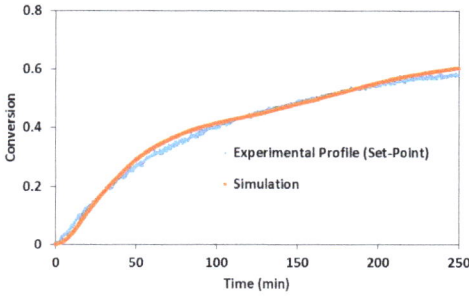

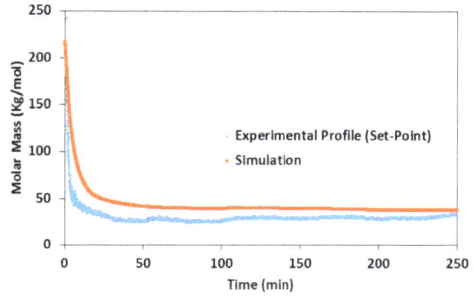

Figure 14. Closed-loop results using single loop controller.

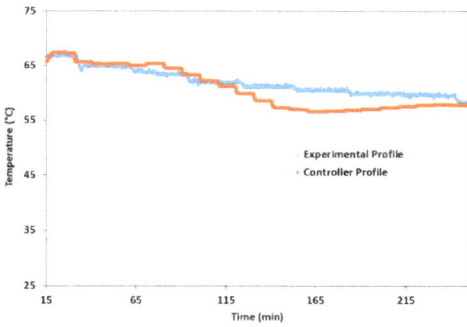

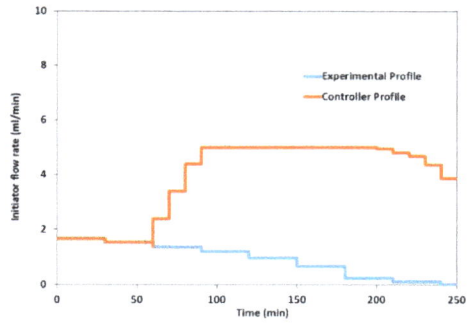

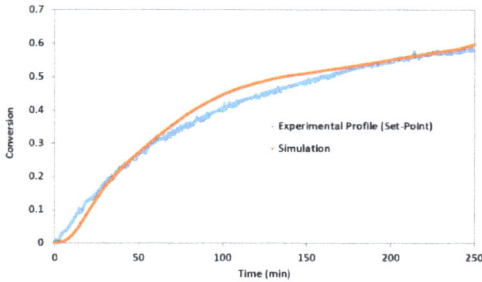

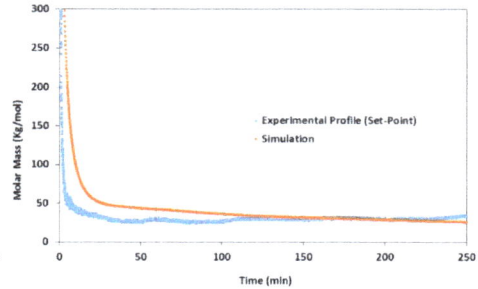

Figure 15. Closed-loop results for multi-loop controllers.

107

5. Conclusions

Formulation and implementation of a model-based framework for prediction and understanding of free radical polymerization processes has been dealt in this paper. The proposed framework is based on the recent development in characterizing molecular weight distribution (MWD) in polymerization processes combined with a state-of–the art experimental tools for on-line monitoring of polymeric properties towards the optimal operations of this type of systems. The method of finite molecular weight moments has been proposed to represent MWD. This allows computation of the complete chain length distribution instead of molecular weight averages. The model presented was modified for application in semi batch processes. The parameters of the kinetic model were then identified and estimated for the case study of MMA free radical polymerization. Dynamic optimization of the system using the CVP technique was undertaken to yield a product with desired polymer molecular weight characteristics in terms of monomer conversion, weight average molecular weight and the entire desired molecular weight distribution. The model-based optimal policy was validated experimentally and fair to good results were achieved. Preliminary feedback control algorithms were implemented and tested through simulations to achieve a desired conversion and weight average molecular weight of the system. They included single-loop and multi-loop Proportinal-Integral control strategies.

Building on these results, current efforts focus on developing the potential and expanding the capabilities of this new on-line measuring method and modelling approach to fully characterize and control the dynamics involved in polymerization systems. Incorporating advanced control policies with an on line execution level controller and application to other polymerization processes will be the subject of the forthcoming paper.

Acknowledgments: The authors acknowledge support from United States Department of Energy, Advanced Manufacturing Office, DE-EE0005776 and the National Science Foundation under the NSF EPSCoR Cooperative Agreement No. EPS-1430280 with additional support from the Louisiana Board of Regents.

Author Contributions: Navid Ghadipasha and Aryan Geraili conceived the programming and design algorithm for modelling, simulation, estimation, optimization and control part under Jose A. Romagnoli's supervision. The experiments were designed and performed by Michael F. Drenski and Carlos A. Castor under Wayne F. Reed's Supervision. The section "Experimental system" was written by Michael F. Drenski. Navid Ghadipasha wrote the other sections while everyone supervised the writing process.

Conflicts of Interest: The authors declare no conflict of interest.

References

1. Wu, T.; Yu, L.; Cao, Y.; Yang, F.; Xiang, M. Effect of molecular weight distribution on rheological, crystallization and mechanical properties of polyethylene-100 pipe resins. *J. Polym. Res.* **2013**, *20*, 1–10.

2. Schimmel, K.H.; Heinrich, G. The influence of the molecular-weight distribution of network chains on the mechanical-properties of polymer networks. *Colloid Polym. Sci.* **1991**, *269*, 1003–1012.

3. Malekmotiei, L.; Samadi-Dooki, A.; Voyiadjis, G.Z. Nanoindentation study of yielding and plasticity of poly(methyl methacrylate). *Macromolecules* **2015**, *48*, 5348–5357.

4. Guyot, A.; Guillot, J.; Pichot, C.; Guerrero, L.R. New design for production of constant composition co-polymers in emulsion polymerization—Comparison with co-polymers produced in batch. *Abstr. Pap. Am. Chem. S* **1980**, *180*, 131–ORPL.

5. Dimitratos, J.; Georgakis, C.; Elaasser, M.S.; Klein, A. Dynamic modeling and state estimation for an emulsion copolymerization reactor. *Comput. Chem. Eng.* **1989**, *13*, 21–33.

6. Dimitratos, J.; Georgakis, C.; Elaasser, M.; Klein, A. An experimental-study of adaptive kalman filtering in emulsion copolymerization. *Chem. Eng. Sci.* **1991**, *46*, 3203–3218.

7. Hammouri, H.; McKenna, T.F.; Othman, S. Applications of nonlinear observers and control: Improving productivity and control of free radical solution copolymerization. *Ind. Eng. Chem. Res.* **1999**, *38*, 4815–4824.

8. Dochain, D.; Pauss, A. Online estimation of microbial specific growth-rates—An illustrative case-study. *Can. J. Chem. Eng.* **1988**, *66*, 626–631.

9. Kozub, D.J.; Macgregor, J.F. State estimation for semibatch polymerization reactors. *Chem. Eng. Sci.* **1992**, *47*, 1047–1062.

10. Mutha, R.K.; Cluett, W.R.; Penlidis, A. On-line nonlinear model-based estimation and control of a polymer reactor. *Aiche. J.* **1997**, *43*, 3042–3058.

11. Mutha, R.K.; Cluett, W.R.; Penlidis, A. A new multirate-measurement-based estimator: Emulsion copolymerization batch reactor case study. *Ind. Eng. Chem. Res.* **1997**, *36*, 1036–1047.

12. Kravaris, C.; Wright, R.A.; Carrier, J.F. Nonlinear controllers for trajectory tracking in batch processes. *Comput. Chem. Eng.* **1989**, *13*, 73–82.

13. Alhamad, B.; Romagnoli, J.A.; Gomes, V.G. On-line multi-variable predictive control of molar mass and particle size distributions in free-radical emulsion copolymerization. *Chem. Eng. Sci.* **2005**, *60*, 6596–6606.

14. Garcia, C.E.; Morari, M. Internal model control.1. A unifying review and some new results. *Ind. Eng. Chem. Proc. Dd.* **1982**, *21*, 308–323.

15. Park, M.J.; Rhee, H.K. Control of copolymer properties in a semibatch methyl methacrylate/methyl acrylate copolymerization reactor by using a learning-based nonlinear model predictive controller. *Ind. Eng. Chem. Res.* **2004**, *43*, 2736–2746.

16. Gattu, G.; Zafiriou, E. Nonlinear quadratic dynamic matrix control with state estimation. *Ind. Eng. Chem. Res.* **1992**, *31*, 1096–1104.

17. Lee, J.H.; Ricker, N.L. Extended kalman filter based nonlinear model-predictive control. *Ind. Eng. Chem. Res.* **1994**, *33*, 1530–1541.

18. Henson, M.A. Nonlinear model predictive control: Current status and future directions. *Comput. Chem. Eng.* **1998**, *23*, 187–202.

19. Schork, F.J.; Deshpande, P.B.; Leffew, W.K. *Control of polymerization reactors*; CRC Press: Boca Raton, FL, USA, 1993; pp. 101–104.

20. Crowley, T.J.; Choi, K.Y. Experimental studies on optimal molecular weight distribution control in a batch-free radical polymerization process. *Chem. Eng. Sci.* **1998**, *53*, 2769–2790.

21. Ellis, M.F.; Taylor, T.W.; Jensen, K.F. Online molecular-weight distribution estimation and control in batch polymerization. *Aiche. J.* **1994**, *40*, 445–462.

22. Congalidis, J.P.; Richards, J.R.; Ray, W.H. Feedforward and feedback-control of a solution copolymerization reactor. *Aiche. J.* **1989**, *35*, 891–907.

23. Adebekun, D.K.; Schork, F.J. Continuous solution polymerization reactor control 2. Estimation and nonlinear reference control during methyl-methacrylate polymerization. *Ind. Eng. Chem. Res.* **1989**, *28*, 1846–1861.

24. Florenzano, F.H.; Strelitzki, R.; Reed, W.F. Absolute, on-line monitoring of molar mass during polymerization reactions. *Macromolecules* **1998**, *31*, 7226–7238.

25. Giz, A.; Catalgil-Giz, H.; Alb, A.; Brousseau, J.L.; Reed, W.F. Kinetics and mechanisms of acrylamide polymerization from absolute, online monitoring of polymerization reaction. *Macromolecules* **2001**, *34*, 1180–1191.

26. Reed, W.F. Automated continuous online monitoring of polymerization reactions (acomp) and related techniques. *Anal. Chem.* **2013**.

27. Baillagou, P.E.; Soong, D.S. Major factors contributing to the nonlinear kinetics of free-radical polymerization. *Chem. Eng. Sci.* **1985**, *40*, 75–86.

28. Pinto, J.C.; Ray, W.H. The dynamic behavior of continuous solution polymerization reactors 7. Experimental-study of a copolymerization reactor. *Chem. Eng. Sci.* **1995**, *50*, 715–736.

29. Ray, W.H. Mathematical modeling of polymerization reactors. *J. Macromol. Sci. R M C* **1972**, *8*, 1–56.

30. Crowley, T.J.; Choi, K.Y. Calculation of molecular weight distribution from molecular weight moments in free radical polymerization. *Ind. Eng. Chem. Res.* **1997**, *36*, 1419–1423.

31. Crowley, T.J.; Choi, K.Y. Optimal control of molecular weight distribution in a batch free radical polymerization process. *Ind. Eng. Chem. Res.* **1997**, *36*, 3676–3684.

32. Ross, R.T.; Laurence, R.L. Gel effect and free volume in the bulk polymerization of methyl methacrylate. *Aiche. J.* **1976**, *72*, 74–79.

33. Fan, S.; Gretton-Watson, S.P.; Steinke, J.H.G.; Alpay, E. Polymerisation of methyl methacrylate in a pilot-scale tubular reactor: Modelling and experimental studies. *Chem. Eng. Sci.* **2003**, *58*, 2479–2490.

34. Li, R.J.; Henson, M.A.; Kurtz, M.J. Selection of model parameters for off-line parameter estimation. *IEEE T Contr. Syst. T* **2004**, *12*, 402–412.

35. Nowee, S.M.; Abbas, A.; Romagnoli, J.A. Optimization in seeded cooling crystallization: A parameter estimation and dynamic optimization study. *Chem. Eng. Process.* **2007**, *46*, 1096–1106.

Surrogate Models for Online Monitoring and Process Troubleshooting of NBR Emulsion Copolymerization

Chandra Mouli R. Madhuranthakam and Alexander Penlidis

Abstract: Chemical processes with complex reaction mechanisms generally lead to dynamic models which, while beneficial for predicting and capturing the detailed process behavior, are not readily amenable for direct use in online applications related to process operation, optimisation, control, and troubleshooting. Surrogate models can help overcome this problem. In this research article, the first part focuses on obtaining surrogate models for emulsion copolymerization of nitrile butadiene rubber (NBR), which is usually produced in a train of continuous stirred tank reactors. The predictions and/or profiles for several performance characteristics such as conversion, number of polymer particles, copolymer composition, and weight-average molecular weight, obtained using surrogate models are compared with those obtained using the detailed mechanistic model. In the second part of this article, optimal flow profiles based on dynamic optimisation using the surrogate models are obtained for the production of NBR emulsions with the objective of minimising the off-specification product generated during grade transitions.

Reprinted from *Processes*. Cite as: Madhuranthakam, C.M.R.; Penlidis, A. Surrogate Models for Online Monitoring and Process Troubleshooting of NBR Emulsion Copolymerization. *Processes* **2016**, *4*, 6.

1. Introduction

Nitrile butadiene rubber (NBR) is an elastomer used in a wide variety of applications demanding oil, fuel and chemical resistance where the content of acrylonitrile influences the end use. NBR can be produced by emulsion copolymerization of acrylonitrile (AN) and butadiene (Bd) using batch, semi-batch, and continuous processes. Usually it is produced using a series of eight to ten continuous-stirred tank reactors (CSTRs). Cold NBR polymers are synthesized between 5 and 15 °C, while hot NBR polymers are usually synthesized between 30 and 50 °C. A comprehensive mechanistic model that can predict different property trajectories for NBR emulsion polymerization has been developed by our group and has been successfully verified with experimental results over a long period [1–3]. This model is capable of simulating the emulsion polymerisation of NBR in batch and in a train of CSTRs, with add-on options, such as choosing the type of reactor start-up, different modes of monomer partitioning, and the effect of impurities.

111

More detailed information regarding the traits and attributes of this model can be obtained elsewhere [4,5]. The mechanistic model developed by our group is complete and comprehensive and has been used for more than just obtaining the simulated dynamic behavior of commercial trains of CSTRs corresponding to different operating conditions. Depending on the type of start-up of the reactor train and different mechanisms selected by the user to be operative (related to radical desorption, partitioning methods, *etc.*), the simulation time varied from several minutes to an hour. In the case of starting up the reactors full of water or empty, the simulation time was found to be almost two hours, depending on the detail of the selected thermodynamic approach for monomer partitioning. To integrate the process model for control and optimisation applications, though, the current mechanistic model can be used, as it adds significant delay in the response of the measured variables before the control action is taken. To overcome this problem, suitable surrogate models (whose order is significantly less than the order of the actual mechanistic model and whose simulation times are far less than that of the fully mechanistic model) are proposed and used in this article for further online applications. The objective of the current article is two-fold: firstly, we explain the need for using surrogate models (in addition to the detailed model developed by our group), obtained using various techniques such as neural networks and transfer function models; secondly, we use them for real-time process applications, such as recipe formulations, control, and dynamic optimisation.

Surrogate models can come to the rescue when the objective is to control or optimise a process whose dynamics are either complex or involves relatively tedious and time-consuming numerical analysis to solve the original complex model for obtaining different state variables, or when the actual physics/chemistry of the process are poorly understood/not known. Artificial neural networks (ANN) are an important and useful tool that belongs to the class of surrogate modeling and is used for control and optimisation of processes which are highly nonlinear [6,7]. Several authors have reported using ANN alone for predicting dynamic behavior or for controlling polymerization reactors [8,9], or otherwise, in a hybrid mode where ANN is used in addition to the corresponding simpler mechanistic model of the system [10,11]. In general, ANNs are highly efficient when trained with large datasets involving a wide range of operating conditions. Otherwise, their performance will be limited and their predictions will be different from the expected dynamics [11]. With sufficient training data, ANN can efficiently be used for prediction of steady state properties as reported by Vijayabhaskar *et al.* [12], Assenhaimer *et al.* [13], and Delfa and Boschetti [14]. For emulsion copolymerisation systems. Although ANNs are used like a black box for modeling processes that are nonlinear, very little is available in the literature for using them as inverse modeling tools for complex polymerisation processes. In the current article, an inverse modeling approach with

ANNs based on the back propagation technique is used for obtaining formulations for recipe ingredients to be used in the first reactor of the train that will give desired properties of the copolymer in the last reactor of the train (in a continuous mode). To further illustrate the points and shed more light, the capability of ANNs to predict the dynamic behaviour of emulsion copolymerisation of NBR in a batch reactor is described and discussed next. A major contribution of this research article (in addition to using ANNs) is to discuss how transfer function models are obtained for emulsion copolymerisation of NBR. For the first time, a complex process like the NBR system is described using transfer functions, which are, in turn, used for controlling the properties of the final product.

2. ANNs for NBR Emulsion Copolymerization in a Batch Reactor

In this section we explain how ANNs are used to simulate the production of NBR emulsion copolymerisation in a batch reactor. With the mechanistic model, the overall simulation time strongly depended on several factors such as the type of method for calculating monomer partitioning (*i.e.*, thermodynamics *vs* constant partition coefficients), presence of monomer or water soluble impurities, and other details in the copolymerization kinetics. For any fast control action to be taken, relying on the mechanistic model would be equivalent to adding a dead time or delay to the measurement signal, the typical fast measurement being on conversion (X), with other rather slow measurements on copolymer composition or particle size or Mooney viscosity. Mooney viscosity is a well-known indirect measure of an average molecular weight of a polymer (usually a rubber product), determined by a Mooney visometer. In batch operation, ANN is designed to predict the effect of time on important latex and polymer properties such as conversion, number of particles, copolymer composition (CPC), weight-based and number-based average molecular weights, and tri- and tetra-functional branching frequencies. In the simulations, ANN based on back-propagation is programmed in MATLAB, where the total prediction error, $E(\underline{W})$ (given by Equation (1)), is minimized. The error, $E(\underline{W})$, also referred to as quadratic error, is defined as the sum of the squares of the differences between the desired output, Y_i, and the predicted output, X_i. The predicted output X_i, is a function of the weights (W_{jk} for one hidden layer) used in the network, where the subscripts j and k represent the indices of the input and output neurons. In vector notation, W_{jk} is usually represented by $\underline{W} = (W_{12}, W_{13}, W_{14}, ...)$:

$$E(\underline{W}) = \frac{1}{2}\sum [Y_i - X_i(\underline{W})]^2 \tag{1}$$

The predicted output is operated upon by an activation function also known as a squashing function. The two most commonly used activation functions are the sigmoid and hyperbolic tangent. It should be noted that, in the absence of

113

the activation function, the problem reduces to that of a multiple linear regression model. It is the activation function that provides the ability to handle nonlinearities. The best values for the weights, W_{jk}, are obtained by training the network using Levenberg–Marquardt back-propagation algorithm. In this algorithm, the weights are adjusted using the method of steepest descent with respect to the error E, as defined by Equation (1). The builtin function, trainlm.m, in MATLAB is based on this algorithm and is used in the simulations for the NBR system. The desired output, Y, is obtained from the mechanistic model, which gives good predictions when compared with the experimental data, as established in previous work [4,5]. The original mechanistic model is highly nonlinear with 32 multiscale state variables. The data obtained from the mechanistic model are divided into training, validation, and untested datasets (also called unseen datasets). From the available data sets, 70% of the data is used for training, 15% percent of the data is used for validation, and the remaining 15% of the data is used for testing. To avoid an overfitted model for the training data, a program in MATLAB is written for obtaining the optimum size and complexity of the network with the objective that the training error be comparable to the prediction error. Performance characteristics of the designed ANN for the prediction of conversion, cumulative copolymer composition (CPC), weight-based average molecular weight (MW_w) and tri-functional branching frequency (BN_3) are shown in Figure 1a–d. In these figures, the profiles obtained from the mechanistic model (MM) are compared to those obtained using ANN for a targeted batch time of 700 min (a typical value used in commercial production). A very good agreement was achieved between the predictions using ANN with those obtained using the well-established and tested MM. The designed ANN for NBR emulsion copolymerisation in a batch reactor can, thus, be safely used in conjunction with process control and optimisation algorithms for describing the desired properties of the polymer.

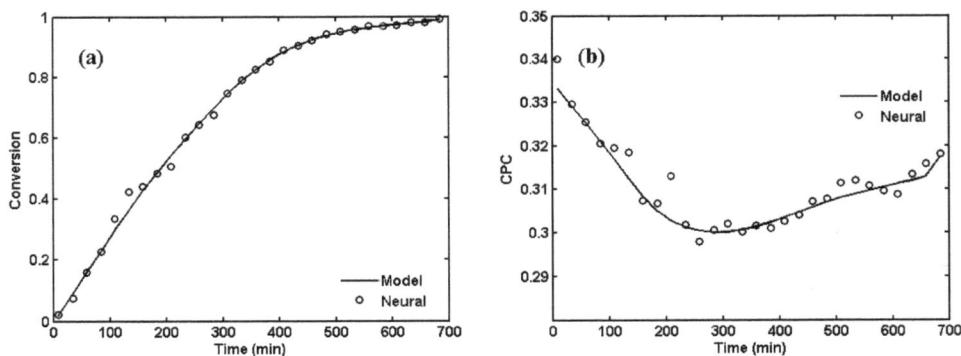

Figure 1. *Cont.*

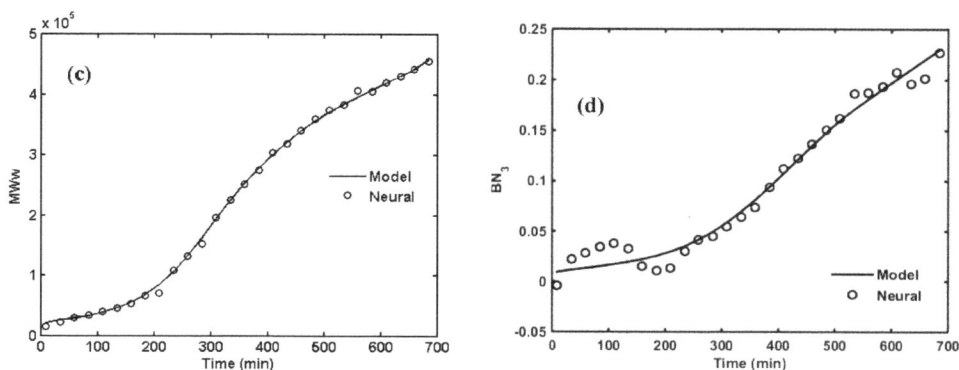

Figure 1. Comparison of (**a**) conversion; (**b**) cumulative copolymer composition; (**c**) weight-average molecular weight; and (**d**) tri-functional branching frequency profiles obtained using ANN and mechanistic models. For process conditions, refer to Dube *et al.* [2].

3. ANN for Inverse Modeling of NBR Emulsion Copolymerisation in a Train of CSTRs

NBR is commercially produced in a continuous fashion using eight to ten reactors operated in series. Depending on the demand, a CSTR can be added or removed from the series, which indirectly affects the mean residence time of the train of CSTRs. The properties of the product from the last reactor are affected by the recipe ingredients used in the first reactor of the train and by the operating conditions of the train. While ANNs can be used to predict these properties based on recipe ingredients given as inputs, by using ANN-based inverse modeling, the recipe ingredients to be used in the first reactor for targeted properties exiting the last reactor can be obtained. This is achieved by using the properties of the product from the last reactor (say, the eighth reactor in an eight-reactor CSTR train) to be the inputs to the network, while the outputs are the recipe ingredients to the first reactor. This type of inverse modeling is easy and efficient with ANNs compared to the alternative, *i.e.*, offline optimisation using the corresponding mechanistic model as a constraint. The ingredients that are fed to the first reactor are initiator (I), reducing agent (RA), emulsifier(s) (E), monomer(s) (M), water (W), and chain transfer agent (CTA). Considering typical high and low levels for each one of these reaction ingredients, the corresponding conversion (X), cumulative copolymer composition (CPC), and the weight-based average molecular weight (MW_w) at the exit of the eighth reactor can be simulated using the mechanistic model. Out of the 64 available datasets, 52 datasets are then used for training the network and 12 datasets are left in order to check the performance of the ANN with respect to inverse modeling of the recipe ingredients for obtaining the desired polymer properties. The different

115

levels of the reaction ingredients used for MM simulations are shown in Table 1. The low and high levels of the reaction ingredients can be normalized to -1 and $+1$, respectively which, in turn, are used as outputs from the network. The inputs to the network, as shown in Figure 2, are X, CPC, and MW_w, while the outputs are the recipe ingredients RA, I, E, M, W, and CTA.

Table 1. Recipe ingredients to the first reactor in the reactor train and their levels.

Ingredient	Low Level (L/min)	High Level (L/min)
Sodium Formaldehyde Sulfoxylate (RA)	0.165	0.22
p-methane hydroperoxide (I)	0.046	0.062
Dresinate/Tamol (E)	0.89/1.67	1.183/2.228
Acrylonitrile/Butadiene (M)	48.6/160.3	64.8/213.7
Water (W)	121.36	161.81
tert-dodecyl Mercaptan (CTA)	0.33	0.44

Note: Mean residence time of each reactor in train is 60 min.

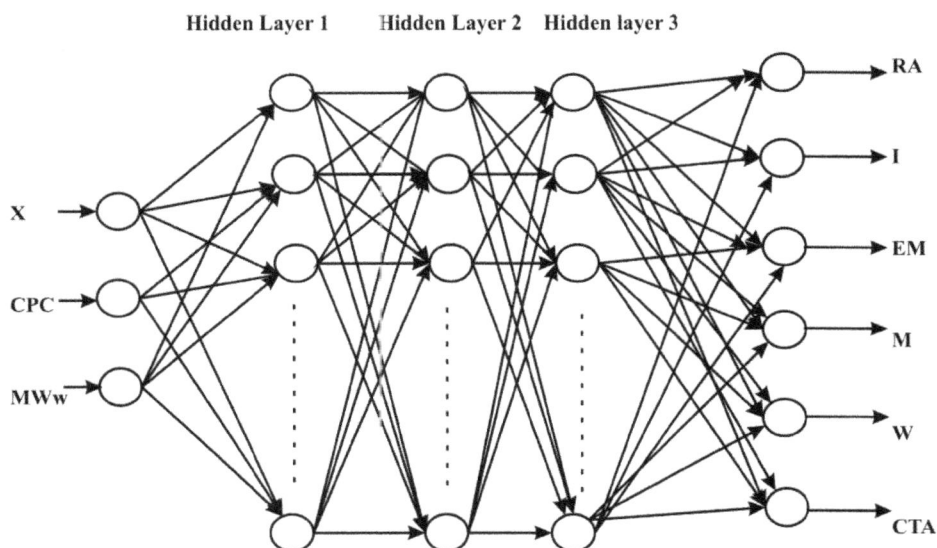

Figure 2. ANN structure used for inverse modeling with X, CPC, and MW_w as inputs; RA, I, E, M, W, and CTA as outputs.

With the objective of obtaining the minimum mean sum of squared errors (MSE), simulations were performed to study the effect of the number of hidden layers and the number of neurons in each layer. Table 2 shows the MSE values for monomer and CTA concentrations obtained by varying the number of hidden layers from one

116

to three and the corresponding number of neurons in each layer from five to 20. All ANN simulations took less than 3 s for obtaining the trained networks. As is evident from Table 2, a network with three hidden layers and 20 neurons gives minimum MSE values and the same trend was found for other reaction ingredients (I, RA, W, and E).

Table 2. Effect of number of hidden layers and number of neurons on MSE.

Number of Neurons	Number of Hidden Layers					
	Monomer			CTA		
	1	2	3	1	2	3
5	0.1532	0.1593	0.0033	0.1370	0.0420	0.0405
10	0.0653	0.0289	0.0836	0.0677	0.0509	0.2600
15	0.0148	0.0659	0.0405	0.0143	0.2070	0.1355
20	0.0245	0.2412	0.0124	0.137	0.1175	0.0613

A possible reason that can be attributed to the resulting large network is that the difference in the magnitude for the output variables such as conversion and weight-based molecular weight is almost 10^6. Hence, the optimum configuration used in this work was a network with three hidden layers with 20 neurons in each layer. Though the number of weights for such a large network is very high, the simulation time for obtaining the trained network in all simulations was only a few seconds. The trained network (saved as .MAT file in MATLAB) in turn is used as an inverse modeling tool to predict the recipe ingredients (of the first reactor) for targeted properties of the stream exiting the eighth reactor. Figure 3 (a through f) shows the recipe ingredients' predictions obtained using the ANN-based inverse modeling for desired conversion (X) compared to the data obtained from the mechanistic model. The proposed ANN-based inverse modeling could predict the required recipe ingredients very well for desired conversion levels. Similar results and trends were obtained for desired CPC and MW_w (as shown in Figures 4 and 5). The predictions for all reaction ingredients are precise, except for the initiator in some cases. The prediction capability of the ANN can be quantified using the mean sum of squared errors (MSE) for each dataset. The MSE values for each of the predicted datasets are shown in Table 3.

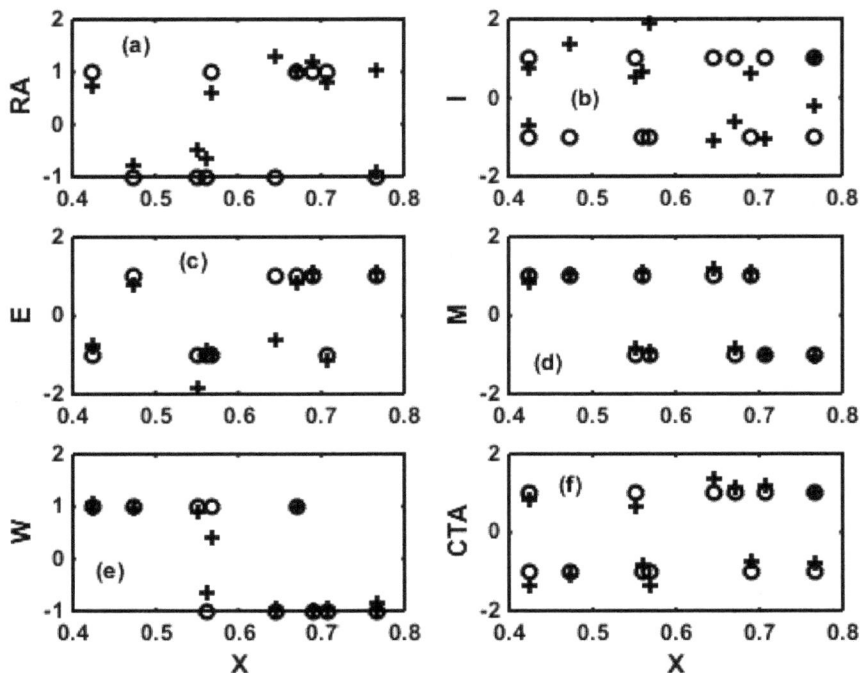

Figure 3. Comparison of the predictions between the unseen targeted values (o) and the values obtained using ANN (+) for desired conversion levels *vs.* (**a**) reducing agent; (**b**) initiator; (**c**) emulsifier; (**d**) monomer; (**e**) water; and (**f**) chain transfer agent.

Table 3. MSE values for the 12 untested datasets using ANN.

#	X	CPC	MW_w	Mean Squared Error (MSE)					
				RA	I	E	M	W	CTA
1	0.5611	0.2573	1.17×10^5	0.1247	2.7766	0.0136	0.0105	0.1115	0.0255
2	0.7668	0.2811	1.51×10^5	0.0096	4.0446	0.0074	0.0042	0.0240	0.0001
3	0.4743	0.2735	8.54×10^4	0.0518	5.6491	0.0384	0.0014	0.0016	0.0068
4	0.5515	0.2285	7.56×10^4	0.2717	0.2051	0.7459	0.0300	0.0110	0.1170
5	0.7669	0.2811	1.95×10^5	4.1140	1.4764	0.0101	0.0015	0.0002	0.0396
6	0.6454	0.2711	1.19×10^5	5.2109	4.4500	2.6381	0.0287	0.0034	0.1231
7	0.5689	0.2349	1.05×10^5	0.1591	8.3361	0.0016	0.0076	0.3575	0.1394
8	0.4246	0.2676	5.45×10^4	0.0674	2.9982	0.0549	0.0037	0.0053	0.0211
9	0.6895	0.2784	1.80×10^5	0.0325	2.6604	0.0100	0.0094	0.0000	0.0626
10	0.7074	0.2713	1.24×10^5	0.0470	4.2123	0.0217	0.0001	0.0024	0.0424
11	0.4247	0.2676	7.22×10^4	0.0743	2.9586	0.0487	0.0206	0.0027	0.1336
12	0.6703	0.2547	1.16×10^5	0.0014	2.6040	0.0243	0.0321	0.0004	0.0246

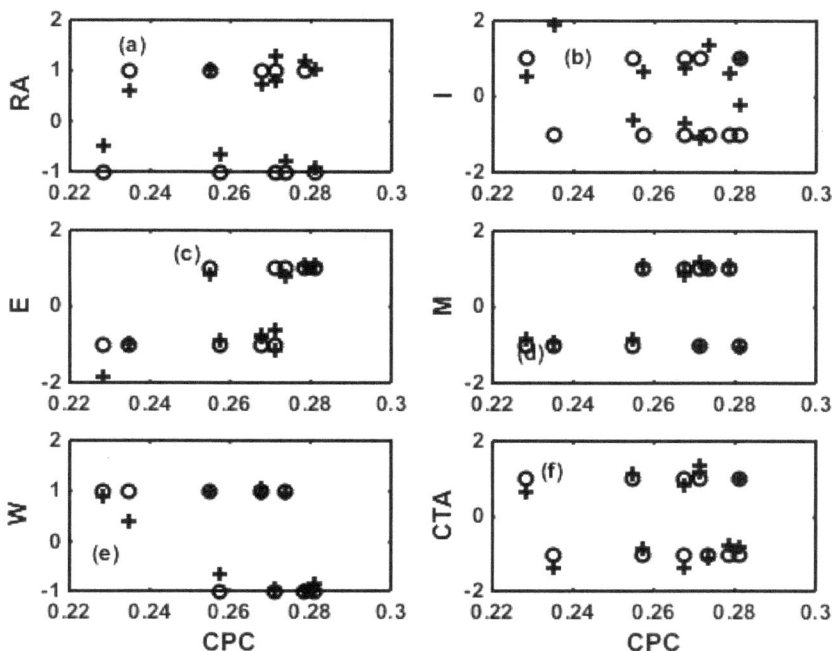

Figure 4. Comparison of the predictions between the unseen targeted values (o) and the values obtained using ANN (+) for desired cumulative copolymer composition levels *vs.* (**a**) reducing agent; (**b**) initiator; (**c**) emulsifier; (**d**) monomer; (**e**) water; and (**f**) chain transfer agent.

The MSE values from Table 3 clearly show that the prediction using ANN-based inverse modeling is precise for almost all reaction ingredients except for the initiator. The obtained the MSE of the initiator is slightly greater than the MSE values of the other reaction ingredients. This is due to the fact that the magnitude of initiator concentration is very small compared to the concentrations of the other ingredients. The above results show that ANN-based inverse modeling can give good estimates of the reaction ingredients to be used in the first reactor (which can be applied to batch reactor operation as well) for obtaining the desired properties of the polymer exiting the last reactor of the CSTR train. This method of obtaining the estimates (for initial reaction ingredients) is easier and less time consuming compared to using the fully mechanistic model. Using the mechanistic model (which is very useful in its own right, as we have shown in previous publications, e.g., Madhuranthakam and Penlidis [4,5]), a trial and error approach has to be used for initial reaction ingredients. During the operation of the reactor train, for any slight discrepancies in the desired properties of the products, it is always possible to fine-tune the estimates obtained using the ANN-based inverse modeling.

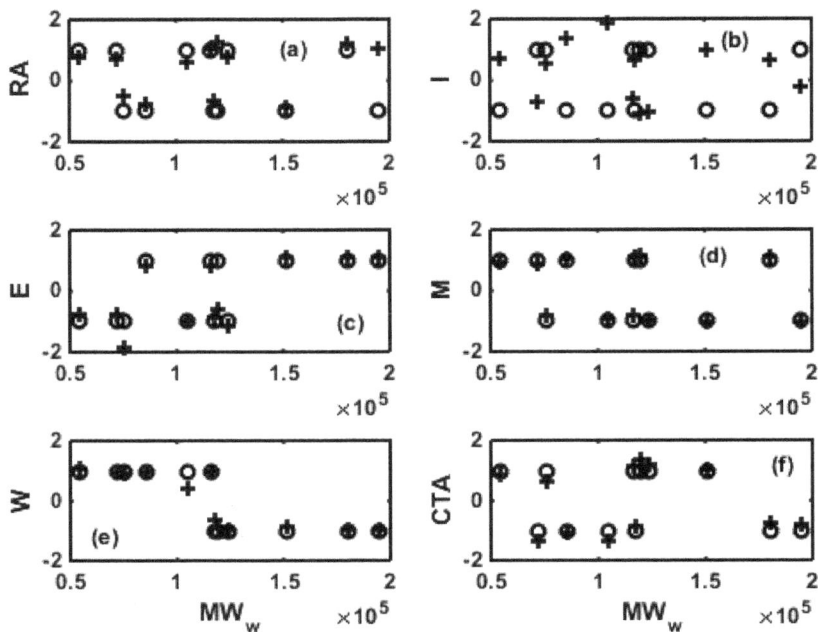

Figure 5. Comparison of the predictions between the unseen targeted values (o) and the values obtained using ANN (+) for desired weight-based average molecular weight levels *vs.* (**a**) reducing agent; (**b**) initiator; (**c**) emulsifier; (**d**) monomer; (**e**) water; and (**f**) chain transfer agent.

4. Surrogate Modeling for NBR Emulsion Copolymerization

In this section, models that are capable of predicting the dynamics and are amenable for control and/or optimisation applications for the emulsion copolymerisation of NBR are discussed. The ANNs discussed in the previous sections can also be used for predicting the dynamics and for control purposes, but the additional benefit of the surrogate models is that these models in their standard forms, such as a first order plus time delay or a second order plus time delay, *etc.*, have fewer parameters than the number of weights obtained using the ANN. In many situations, the parameters of a feedback controller or a model predictive controller can be obtained as functions of the corresponding parameters of the model, which is not feasible with the case of ANN. Surrogate models are obtained by reducing the order of the original mechanistic model so that computation of the dynamic behaviour is fast, which in turn helps with online control and optimisation applications. Depending on the type and order of the original model, the model can be reduced either by using a model balancing approach or by error minimization. Model balancing involves evaluating the controllability and observability Gramians and partitioning the state vector into important states and less important states. The reduced model is obtained by truncation of the least important states [15]. This method involves linearizing the original model around an operating condition or empirically obtaining Gramians corresponding to each operating condition, which may be very tedious. The actual mechanistic model for NBR emulsion copolymerization includes 32 state variables and its highly nonlinear nature makes it cumbersome to obtain a corresponding linearized model. Due to this reason, empirical models are obtained by using error minimization criteria. The initial choice of type of empirical models is very crucial as there could be multiple models (which could differ in the number of parameters) that can fit equally well the corresponding data available. After choosing a specific transfer function model, the models are fine-tuned later based on the objective of minimizing the error between the responses of the proposed empirical model and the data (obtained from the mechanistic model). The parameters of the final surrogate model are obtained by simulation and using the nonlinear least squares fitting function lsqnonlin in MATLAB.

5. Transfer Function Models for the First CSTR in the Reactor Train

As mentioned in the previous sections, the ingredients entering the first CSTR are the initiator, emulsifier(s), monomers, water, and chain transfer agent streams. Typical outputs considered for obtaining the corresponding surrogate models are conversion, cumulative copolymer composition, weight-based average molecular weight, and the total number of latex particles per liter of water (N_p). These are the typical outputs (in principle, measurable) which are, in turn, used in the controlled production of NBR latex. Surrogate modeling is conducted in the Laplace domain by programming interactive simulations between SIMULINK and MATLAB. The performance of the corresponding fitted transfer function model is evaluated in terms of the coefficient of determination, R^2. The input variables chosen to be related to any output variable are restricted to the states that have higher impact than others and that can also be used as practical manipulated variables in control applications. For example, when the reactor is started full of batch recipe (for other types of start-up policies refer to [4]), the conversion obtained at the exit of the reactor is expressed as a function of initiator, and acrylonitrile and butadiene (monomers) flow rates, as shown in Equation (2):

$$Y_1(s) = \frac{\frac{\tau_2}{K_1}s}{\left(\sqrt{\frac{\tau_1 \tau_2}{K_1}}\right)^2 s^2 + ((1+K_1)\tau_2)s + 1} X_1(s) + \frac{K_1}{\tau_2 s + 1} X_2(s) + \frac{K_1}{\tau_2 s + 1} X_3(s) \quad (2)$$

where Y_1 denotes conversion, X_1 represents initiator flowrate, X_2 is the acrylonitrile flow rate, and X_3 is the butadiene flow rate, all in the Laplace domain. τ_1, τ_2, and K_1

are the parameters of the model. Since the input and output variables in the transfer function models represent perturbations from initial steady states, the final model constitutes an initial value problem with all variables (outputs) to be zero at time $t = 0$. The structure of this model basically consists of a combination of a second order and two first order systems. The parameters τ_1, τ_2, and K are obtained by fitting the model response to the data obtained from the mechanistic model for a given step change in the input variables X_1, X_2, and X_3. Similarly, for other output variables such as CPC (Y_2), MW$_w$ (Y_3), and N_p (Y_4), the corresponding models are given by Equations (3)–(5):

$$Y_2(s) = \frac{K_2}{\tau_3 s + 1} \exp(-\tau_4 s) \, X_4(s) \tag{3}$$

$$Y_3(s) = \frac{K_3 s + 1}{\tau_5^2 s^2 + 2\tau_5 \tau_6 s + 1} X_5(s) \tag{4}$$

$$Y_4(s) = \frac{K_4 s}{\tau_7^2 s^2 + 2\tau_7 \tau_8 s + 1} X_1(s) + \left[\frac{1}{\tau_7 s + 1}\right]^2 X_6(s) \tag{5}$$

where Y_2, Y_3 and Y_4 represent the output variables CPC, MW$_w$, and N_p, respectively, whereas X_4, X_5, and X_6 represent the ratio of flow rates of the two monomers (AN to Bd), chain transfer agent flow rate, and emulsifier flow rate, respectively. K_2, K_3, K_4, and τ_3 through τ_8 are parameters of the empirical models. Equations (2)–(5) represent the empirical surrogate models for the dynamics of the first reactor in the reactor train. When the reactor is started full of recipe, the inflow to the first reactor is equivalent to giving a step input to all input variables X_1 through X_6 and the output responses Y_1 through Y_4 are used to obtain the corresponding parameters of the empirical models. The final comparison of the responses obtained for X, CPC, MW$_w$, and N_p using the data from the mechanistic model (MM) and the proposed empirical models (EM) are shown in Figure 6a–d, respectively.

The proposed empirical models fit very well the data obtained from the mechanistic model is evident from the very high R^2 values (close to unity) reported on the corresponding figures. Since the weight-based average molecular weight (MW$_w$) in the mechanistic model is obtained from a set of very highly nonlinear model equations that originate using the method of moments, the corresponding empirical model had a lower R^2 value compared to the values of the other output variables. In general, the performance of the surrogate models is very good. The proposed empirical models are not only simple and amenable to use for online purposes but also have very few parameters. With the success of using this method for a single CSTR (the first reactor of the CSTR train), the properties at the exit of the eighth reactor can also be empirically modeled and further used in online applications, such as control and grade transitions.

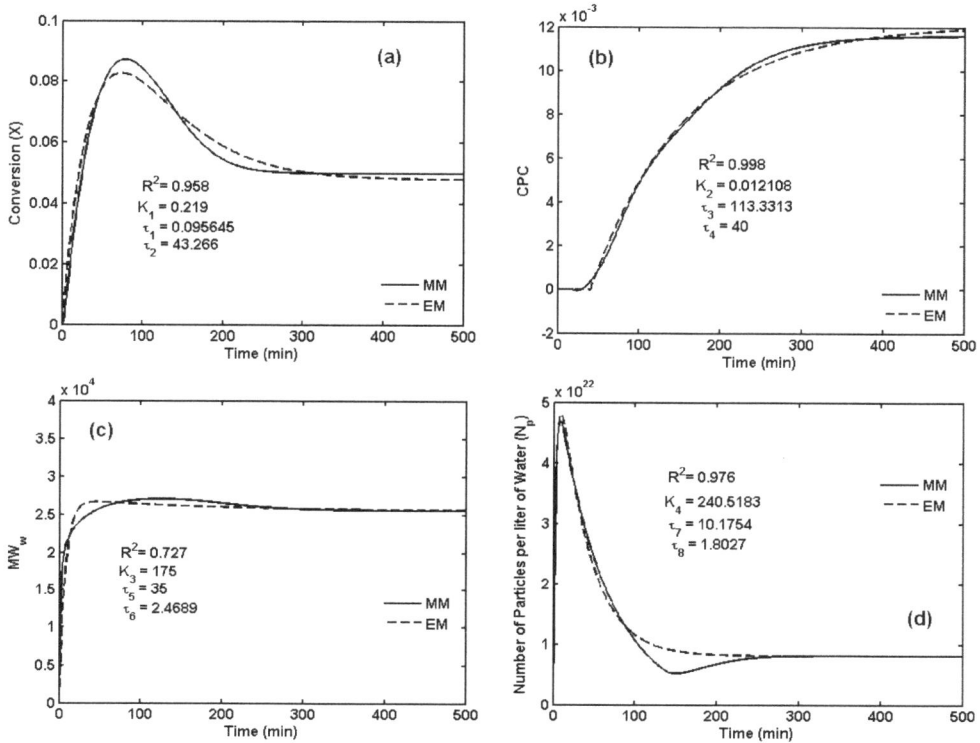

Figure 6. Comparison of the responses obtained from the proposed empirical models (EM) and mechanistic model (MM) for (**a**) conversion; (**b**) CPC; (**c**) MW_w; and (**d**) N_p.

6. Optimal CTA Profile for Minimizing off-Spec Product

The weight-based molecular weight response exiting the eighth reactor in the reactor train is empirically modeled using the knowledge of the corresponding dynamics in the first reactor of the train. The validity and versatility of the proposed empirical model is shown in Figure 7 where two different reactor start-up procedures are used (one of them starting the reactor train full of recipe (Figure 7a) and the other starting full of water (Figure 7b). The empirical model obtained for this case is given by Equation (6).

$$Y_8(s) = \left[\frac{Ks + 1}{\tau_1^2 s^2 + 2\tau_1 \tau_2 s + 1} \right]^8 X_5(s) \tag{6}$$

123

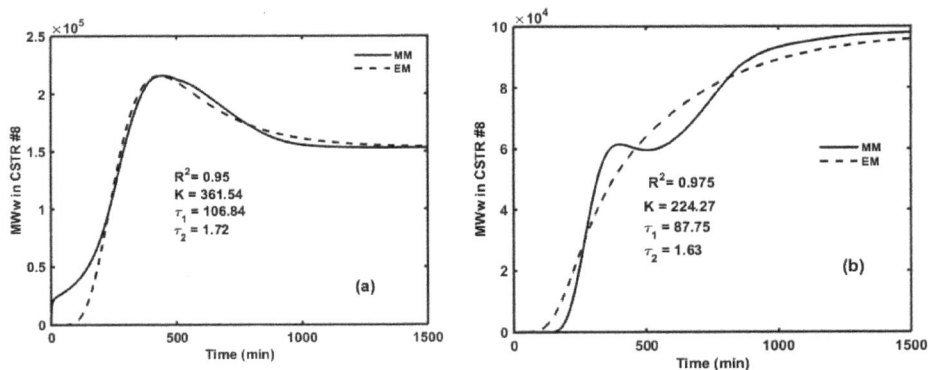

Figure 7. Comparison of the model validity for MW_w at the exit of the eighth reactor using mechanistic model (MM) and empirical model (EM) for reactor train start-ups (**a**) full of recipe and (**b**) full of water.

Y_8 is the MW_w obtained at the exit of the eighth reactor, X_5 is the flow rate of the CTA to the first reactor in the reactor train, and K, τ_1, and τ_2 are model parameters. The empirical model consists of eight second-order transfer functions in series, each of them corresponding to the dynamics of each individual reactor in the reactor train. From the R^2 values reported in Figure 7a,b, it is evident that the proposed second order transfer function in series fits very well the corresponding mechanistic model in addition to the benefit of employing three parameters only (K, τ_1, and τ_2).

One of the most common operational constraints that occur in the commercial production of NBR is during grade changes. During grade changes, the reactor train is switched to operate from one steady state to another desired steady state. These can be viewed as set point changes that occur due to customer or production campaign requirements. One of the primary objectives during grade changes or for start-up of the reactor train is to minimize the off-specification product that is produced before reaching the steady state. The off-specification product can be minimized in multiple ways but a usual practice is to add intermittent flows of the manipulated variables along the reactor train in a feedforward fashion where the magnitudes of the flow rates of the manipulated variables are estimated by performing offline optimisation. For batch and semi-batch production of polymers, Fujisawa and Penlidis [16] have shown different reactor control policies targeted for obtaining desired copolymer compositions using an offline mechanistic model. For reducing transients during grade changes for NBR and styrene butadiene systems, Minari et al. [17,18] have used a bang-bang method (which also uses an offline mechanistic model), where predetermined quantities of CTA and monomer are added along the reactor train in one shot to reduce the molecular weights and the CPC in the last reactor of the reactor train. In the present work we propose an online method, where continuous profiles for manipulated variables such as CTA, monomers *etc.* are obtained using dynamic optimisation. This method assumes that the information for molecular weight becomes available online via an inferential estimator based on Mooney viscosity, in order to minimize the time delay related to the measurement of molecular weight. The empirical model (as given be Equation (6)) can be used either online or offline optimisation when there is a grade change with respect to MW_w. For simultaneous control of conversion, CPC, and MW_w, a multiobjective function based on a weighted sum or ε-constrained methods can be used [19,20]. While grade changes can involve an increase or decrease in several product specifications, such as conversion, CPC, N_P, MW_w, *etc.*, the application to the scenario where a decrease in MW_w by adding extra CTA to the reactor train is discussed here as an example. In general, the inflows to the last few reactors are manipulated rather than manipulating the inflows to the first few reactors, due to the fact that the monomer droplets are absent in the last few reactors. For example, in a reactor train started up full of recipe, the monomer droplets will disappear in the sixth reactor of the train [5]. Especially for grade changes involving MW_w, the corresponding manipulated variable is the flow rate of CTA added to the reactors with monomer droplets present. The optimal flow rate of CTA to be added to the first reactor in the reactor train is obtained using the optimisation function represented by Equation (7):

$$
\begin{aligned}
\min_{F_{CTA}} \; f &= \left| MW_w^{des} - MW_w(t) \right| \\
&subject\ to: \\
&0 \leqslant t \leqslant 1500 \\
&\frac{1}{7} F_{CTA}^* \leqslant F_{CTA} \leqslant 7 F_{CTA}^*
\end{aligned}
\tag{7}
$$

where F_{CAT} is the flow rate of CTA to the eighth reactor, MW_w^{des} is the desired steady state weight-based molecular weight, $MW_w(t)$ is the measured value of the weight-based molecular weight at any time t, and F_{CTA}^* is the steady state value of the flow rate of CTA. Assuming a control valve with a rangeability of 50:1 is used to manipulate the CTA flow rate, the manipulated flow rate is constrained between $\frac{1}{7} F_{CTA}^*$ and $7 F_{CTA}^*$. In Equation (7), time t refers to the operation time of the eighth reactor. The mean residence time for each CSTR in the train is 60 min; hence, the total time for a reactor train of eight CSTRs will be 480 min. Since it takes three times the total mean residence time for the reactor train to reach steady state operation, the corresponding operational time used in the simulations was set to 1500 min (approximately). The optimum value for F_{CTA} is obtained by minimizing the cost function (as shown in Equation (7)) at different time steps simultaneously. This procedure can be extended for other grade change applications with specifications on other variables, such as CPC or X, with manipulated variables being the flow rates of monomers, initiator, and/or emulsifiers.

Figure 8a shows the profiles obtained for MW_w at the exit of the eighth reactor for the cases where a regular CTA flow rate based on a "full of recipe" start-up is compared to that of the CTA flow rate obtained from optimisation using Equation (7). In both cases (refer to Figure 8a), the area under the solid curve and the dashed curve with respect to the steady state value of MW_w is an indirect measure of the amount of off-specification product generated during the operation of the reactor train. Figure 8a clearly shows that using the proposed optimisation method, the amount of off-specification product/material can be minimized by several folds compared to the base case where a constant CTA flow rate is used. The corresponding CTA flow rate profile to be added to the first reactor obtained from the above mentioned optimisation procedure is shown in Figure 8b. This CTA flow profile can be practically achieved by using an automatic flow controller installed on the CTA flow line. The proposed online method for the adjustment of the manipulated CTA flow rate can also be applied to the flow rates of monomers and initiator to control CPC and/or X.

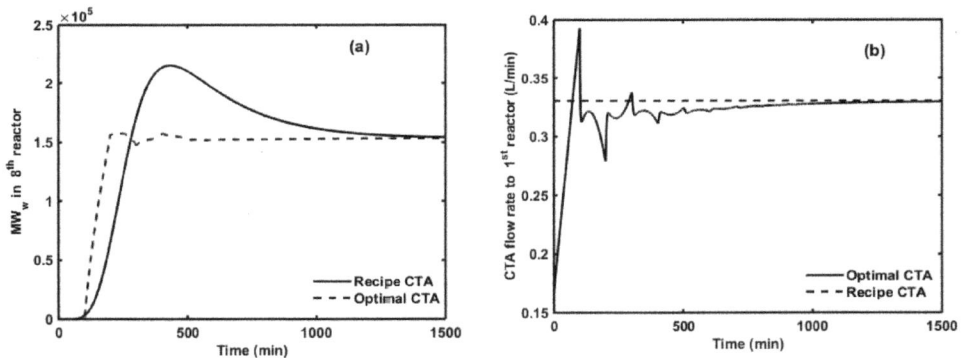

Figure 8. (a) Comparison of MW_w profiles in the eighth reactor using regular CTA flow rate to that of using optimal flow rate of CTA; and (b) the dynamic CTA flow rate obtained from optimisation.

7. Concluding Remarks

Surrogate models were investigated in lieu of the original higher order mechanistic model for NBR emulsion copolymerisation, with the objective of minimizing the computational time for implementing the models for control/optimisation approaches. Different types of surrogate models, such as models based on artificial intelligence using neural networks and empirical models in the form of first order and/or second order (with and without time delay), were designed for studying the dynamics of NBR production. It was shown that ANNs can be used to efficiently predict the dynamics, and also as an inverse modeling tool where the reaction ingredients to be added to the first reactor in the reactor train are obtained for targeted desired properties of the polymer produced in the eighth reactor of the reactor train. The transfer function models were in the form of standard first and second order processes (with or without time delay) and could readily be used in control and optimisation applications. These proposed models fitted well the dynamics of the NBR emulsion polymerization in the CSTR train and were subsequently used in an optimisation application. With the objective of minimizing the off-specification product exiting the eighth reactor, the optimal CTA flow rate was obtained. Compared to offline methods, the proposed (potentially online) method is a very promising tool with respect to optimal reactor train operation and minimizing the waste generated due to different startups or grade changes.

Acknowledgments: The authors wish to acknowledge financial support from the Natural Sciences and Engineering Research Council (NSERC) of Canada, and the Canada Research Chair (CRC) program.

Author Contributions: This work is equally contributed by both Chandra Mouli.R. Madhuranthakam and Alexander Penlidis.

Conflicts of Interest: The authors declare no conflict of interest.

Abbreviations

The following abbreviations are used in this manuscript:

NBR	Nitrile Butadiene Rubber
AN	Acrylonitrile
Bd	Butadiene
CSTR	Continuous-Stirred Tank Reactor
ANN	Artificial Neural Network
MM	Mechanistic Model
EM	Empirical Model

References

1. Washington, I.D.; Duever, T.D.; Penlidis, A. Mathematical modeling of acrylonitrile-butadiene emulsion polymerization: Model development and validation. *J. Macromol. Sci. A Pure Appl. Chem.* **2010**, *47*, 747–769.
2. Dube', M.A.; Penlidis, A.; Mutha, R.K.; Cluett, W.R. Mathematical modeling of emulsion copolymerization of acrylonitrile/butadiene. *Ind. Eng. Chem. Res.* **1996**, *35*, 4434–4448.
3. Scott, A.J.; Nabifar, A.; Madhuranthakam, C.R.; Penlidis, A. Bayesian design of experiments applied to a complex polymerization system: Nitrile butadiene rubber production in a train of CSTRs. *Macromol. Theory Simul.* **2015**, *24*, 13–27.

4. Madhuranthakam, C.R.; Penlidis, A. Modeling uses and analysis of production scenarios for acrylonitrile-butadiene (NBR) emulsions. *Polym. Eng. Sci.* **2011**, *51*, 1909–1918.

5. Madhuranthakam, C.R.; Penlidis, A. Improved operating scenarios for the production of acrylonitrile-butadiene emulsions. *Polym. Eng. Sci.* **2013**, *53*, 9–20.

6. Bhat, N.V.; McAvoy, T.J. Determining model structure for neural models by network stripping. *Comput. Chem. Eng.* **1992**, *16*, 271–281.

7. Nascimento, C.A.; Giudici, R.; Guardani, R. Neural network based approach for optimization of industrial chemical processes. *Comput. Chem. Eng.* **2000**, *24*, 2303–2314.

8. Ekpo, E.E.; Mujtaba, I.M. Evaluation of neural networks-based controllers in batch polymerisation of methyl methacrylate. *Neurocomputing* **2008**, *71*, 1401–1412.

9. Lightbody, G.; Irwin, G.W.; Taylor, A.; Kelly, K.; McCormick, J. Neural network modeling of a polymerization reactor. *Proc. IEEE Int. Conf. Control.* **1994**, *1*, 237–242.

10. D'Anjou, A.; Torrealdea, F.J.; Leiza, J.R.; Asua, J.M.; Arzamendi, G. Model reduction in emulsion polymerization using hybrid first-principles/artificial neural network models. *Macromol. Theory Simul.* **2003**, *12*, 42–56.

11. Arzamendi, G.; d'Anjou, A.; Grana, M.; Leiza, J.R.; Asua, J.M. Model reduction in emulsion polymerization using hybrid first-principles/artificial neural network models 2ª long chain branching kinetics. *Macromol. Theory Simul.* **2005**, *14*, 125–132.

12. Vijayabaskar, V.; Gupta, R.; Chakrabarti, P.P.; Bhowmick, A.K. Prediction of properties of rubber by using artificial neural networks. *J. Appl. Polym. Sci.* **2006**, *100*, 2227–2237.

13. Assenhaimer, C.; Machado, L.J.; Glasse, B.; Fritsching, U.; Guardani, R. Use of a spectroscopic sensor to monitor droplet size distribution in emulsions using neural networks. *Can. J. Chem. Eng.* **2014**, *92*, 318–323.

14. Delfa, G.M.; Boschetti, C.E. Optimization of the chain transfer agent incremental addition in SBR emulsion polymerization. *J. Appl. Polym. Sci.* **2012**, *124*, 3468–3477.

15. Hahn, J.; Edgar, T.F. An improved method for nonlinear model reduction using balancing of empirical gramians. *Comput. Chem. Eng.* **2002**, *26*, 1379–1397.

16. Fujisawa, T.; Penlidis, A. Copolymer composition control colicies: characteristics and applications. *J. Macromol. Sci. A Pure Appl. Chem.* **2008**, *45*, 115–132.

17. Minari, R.J.; Gugliotta, L.M.; Vega, J.R.; Meira, G.R. Continuous emulsion styrene-butadiene rubber (SBR) process: Computer simulation study for increasing production and for reducing transients between steady states. *Ind. Eng. Chem. Res.* **2006**, *45*, 245–257.

18. Minari, R.J.; Gugliotta, L.M.; Vega, J.R.; Meira, G.R. Continuous emulsion copolymerization of acrylonitrile and butadiene: Simulation study for reducing transients during changes of grade. *Ind. Eng. Chem. Res.* **2007**, *46*, 7677–7683.

19. Rivera-Toledo, M.; Flores-Tlacuahuac, A. A multiobjective dynamic optimization approach for a methyl-methacrylate plastic sheet reactor. *Macromol. React. Eng.* **2014**, *8*, 358–373.

20. Camargo, M.; Morel, L.; Fonteix, C.; Hoppe, S.; Hu, G.; Renaud, J. Development of new concepts for the control of polymerization processes: Multiobjective optimization and decision engineering. II. Application of a choquet integral to an emulsion copolymerization process. *J. Appl. Polym. Sci.* **2011**, *120*, 3421–3434.

Gaussian Mixture Model-Based Ensemble Kalman Filtering for State and Parameter Estimation for a PMMA Process

Ruoxia Li, Vinay Prasad and Biao Huang

Abstract: Polymer processes often contain state variables whose distributions are multimodal; in addition, the models for these processes are often complex and nonlinear with uncertain parameters. This presents a challenge for Kalman-based state estimators such as the ensemble Kalman filter. We develop an estimator based on a Gaussian mixture model (GMM) coupled with the ensemble Kalman filter (EnKF) specifically for estimation with multimodal state distributions. The expectation maximization algorithm is used for clustering in the Gaussian mixture model. The performance of the GMM-based EnKF is compared to that of the EnKF and the particle filter (PF) through simulations of a polymethyl methacrylate process, and it is seen that it clearly outperforms the other estimators both in state and parameter estimation. While the PF is also able to handle nonlinearity and multimodality, its lack of robustness to model-plant mismatch affects its performance significantly.

Reprinted from *Processes*. Cite as: Li, R.; Prasad, V.; Huang, B. Gaussian Mixture Model-Based Ensemble Kalman Filtering for State and Parameter Estimation for a PMMA Process. *Processes* **2016**, *4*, 9.

1. Introduction

Polymerization reactors offer unique challenges for process modeling, monitoring, and control. The production of polymers of different grades means that the process conditions are changed relatively often. Product quality specifications (usually expressed in terms of constraints on the properties of the molecular weight distribution) and dynamic operation lead to the need for on-line monitoring and control, which further require accurate process models and real-time estimation of states and parameters of the system. Over the years, the most popular estimator used in nonlinear chemical processes—both in general and specifically for polymerization reactors, too—is the extended Kalman filter (EKF) (e.g., [1–8]). However, this estimator involves linearization of the original model at each step, and can be inaccurate for highly nonlinear systems. Our focus in this work is on particle-based estimators, which are derivative free estimators using different sampling methods to generate an ensemble of particles to represent the distributions of the dynamic states of the system.

The most commonly used estimators based on the use of an ensemble of particles are the ensemble Kalman filter (EnKF) [9], the unscented Kalman filter (UKF) [10,11] and the particle filter (PF) [12]. While the EnKF and the UKF provide only the mean and variance of the posterior distribution of the states (since they use a Gaussian assumption for the distributions), the PF, which works on Bayesian principles, can provide estimates for the full distribution of the states even in situations where the distribution is not Gaussian (which occurs in nonlinear systems) by using a set of particles associated with different weights. In practice, the application of the PF to chemical processes is very recent. Chen *et al.* [13] compared the performance of the auxiliary particle filter with an EKF for a batch polymethyl methacrylate process to show that it outperformed the EKF in terms of the root mean squared error for state and parameter estimation. Shenoy *et al.* [14] compared the UKF, EKF, and PF in a case study on a polyethylene reactor simulation to demonstrate that the PF provided more accurate estimation results, but was less robust to plant-model mismatch. Shao *et al.* [15] compared the performance of the PF, EKF, UKF, and moving horizon estimation for constrained state estimation and showed that the constrained PF provides more accurate estimation results compared to other methods.

An important issue with the PF relates to its performance for high dimensional systems. The ensemble Kalman filter (EnKF), on the other hand, has the advantage of being scalable to high-dimensional systems without a prohibitive increase in the size of the ensemble required; however, as stated earlier, the algorithm is based on the assumption that both the prior and posterior distribution of the states can be approximated by the Gaussian distribution, and it may be unreliable when this assumption is not valid.

Polymerization processes can be of high dimension when they are described using population balance models [16,17] and a multimodal distribution of properties such as the particle size and molecular weight, may be desirable [18–20]. This, especially in the presence of model-plant mismatch, creates challenges for both the EnKF and the PF. Also, the nonlinearity of the systems may lead to multimodality in the state distributions.

Recently, the Gaussian mixture model (GMM) has been combined with the ensemble Kalman filter to create a new category of estimators: Gaussian mixture filters. Bengtsson *et al.* [21] proposed the GMM to approximate the prior distribution of the states, but the means and variances of the GMM were approximated directly from the ensemble. In [22], Smith proposed the expectation maximization (EM) algorithm to learn the parameters of the prior distribution modeled by the GMM. In the update step, the idea of Kalman-based filtering was extended to the multimodal scenario; however, the posterior distribution is constrained to be a Gaussian distribution. Dovera and Della Rossa [23] used a different update technique and retained the posterior distribution as a GMM.

In this work, we propose an estimator that belongs to the category of Gaussian mixture filters and provides a full state distribution at each time step that is approximated by the GMM. We extend the idea of the EnKF to priors with multimodal features that are described by the GMM. We present results on the application of this estimator to a polymethyl methacrylate (PMMA) process and compare its performance to that of the EnKF and the PF.

2. State Estimation Techniques for Nonlinear Systems

Consider a dynamic nonlinear system represented by:

$$\left\{ \begin{array}{c} x_n = f\left(x_{n-1}, u_{n-1}, \theta\right) + v_n \\ y_n = H x_n + e_n \end{array} \right. \tag{1}$$

where x_n are the hidden states. u_n and y_n are the inputs and outputs of the system. θ represents the parameters in the model. v_n and e_n are process noise and measurement noise respectively.

In this section, we will introduce the particle filter and the ensemble Kalman filter for these systems, and then describe the GMM-based ensemble Kalman filter that we propose to employ. The performance of the three estimators will be compared for the PMMA system in later sections.

2.1. Particle Filter (PF)

The PF employs a sequential Monte Carlo method that uses a set of sampling techniques to generate samples from a sequence of probability distribution functions.

The particle filter approximates the posterior probability $p\left(x_n|y_n\right)$ with a set of N_s particles $\{x_n^{(i)}\}$. Each particle is assigned a weight $w_n^{(i)}$ and the sum of all weights is unity. Since the probability distribution of the states conditioned on the measurements of the outputs, $p\left(x_n|y_{1:n}\right)$, is usually unknown, these particles are drawn from the importance distribution $q(x_n \,|\, y_{1:n})$. The posterior distribution is given by:

$$p\left(x_n|y_{1:n}\right) = \sum_{i=1}^{N_s} w_n^{(i)} \delta(x_n - x_n^{(i)}) \tag{2}$$

where the recursive update of the weights $w_n^{(i)}$ is given by:

$$w_n^{(i)} = w_{n-1}^{(i)} \frac{p\left(x_n^{(i)} \middle| x_{n-1}^{(i)}\right)}{q\left(x_n^{(i)} \middle| x_{n-1}^{(i)}, y_n\right)} p(y_n|x_n^{(i)}) \tag{3}$$

In the sequential importance resampling (SIR) version of the PF, we choose $q(x_n^{(i)}|x_{n-1}^{(i)}) = p(x_n^{(i)}|x_{n-1}^{(i)})$, so that $w_n^{(i)} = w_{n-1}^{(i)} p(y_n|x_n^{(i)})$, i.e., we draw particles directly from the prior distribution at time instant n.

The N_s particles at time step (n-1) are forwarded through the state transition equation $x_{n|n-1}^{(i)} = f(x_{n-1|n-1}^{(i)}, u_{n-1}, v_{n-1}^{(i)})$ to get a new series of particles $\{x_{n|n-1}^{(i)}\}_{i=1}^{N_s}$ to approximate the prior density $p(x_n|y_{1:n-1})$ at time instant n. The weight $w_n^{(i)}$ associated with each particle is calculated using Equation (3). Then, a resampling step is performed on the prior particles $\{x_{n|n-1}^{(i)}\}_{i=1}^{N_s}$ based on their weights $w_n^{(i)}$ to generate the posterior particles $\{x_{n|n}^{(i)}\}_{i=1}^{N_s}$ such that the weights of all the posterior particles are set to be equal. The full state distribution and its properties can be calculated from the posterior particles.

2.2. Ensemble Kalman Filter (EnKF)

The EnKF was first proposed as a data assimilation technique for highly nonlinear ocean models by Evensen [9] and is a Monte Carlo sampling based variant of the Kalman filter. Like the PF, it also uses an ensemble of particles from which the statistical information of the distribution of the states can be calculated, but it uses the Kalman update. In order to have an explicit analytical expression for the Kalman gain, both the prior and posterior distributions are approximated by the Gaussian distribution. The framework of this algorithm is as follows:

At time step k, N_e particles are drawn from the prior distribution to form the prior ensemble $\{x_{n-1|n-1}^i\}_{i=1,...,N_e}$. In the prediction step, each member of the ensemble $x_{n-1|n-1}^i$ is forwarded through the state transition equation $x_{n|n-1}^i = f(x_{n-1|n-1}^i, u_{n-1}, v_{n-1}^i)$ to get its predicted value, thus forming a predicted ensemble $\{x_{n|n-1}^i\}_{i=1,...,N_e}$. Corresponding to each member of the ensemble, a predicted observation value is obtained; this can be achieved by perturbing the measurement of the output with random measurement error. Let $\{\hat{y}_{n|n-1}^i\}_{i=1,...,N_e}$ denote the predicted observation data.

In the update step, two error matrices are calculated. The error matrix of the predicted state ensemble is defined as:

$$e_{n|n-1}^i = x_{n|n-1}^i - \mu_{n|n-1}^x \tag{4}$$

where $\mu_{n|n-1}^x = \frac{1}{N_e} \sum_{i=1}^{N_e} x_{n|n-1}^i$.

The error matrix of the predicted measurement ensemble is defined as:

$$\varepsilon_{n|n-1}^i = \hat{y}_{n|n-1}^i - \mu_{n|n-1}^y \tag{5}$$

where $\mu_{n|n-1}^{y} = \dfrac{1}{N_e} \sum\limits_{i=1}^{N_e} \hat{y}_{n|n-1}^{i}$.

The cross-covariance between the state prediction ensemble and measurement ensemble is given in Equation (6), and the covariance matrix of the measurement ensemble is given in Equation (7).

$$P_{n|n-1}^{e,\varepsilon} = \frac{1}{N_e - 1} \sum_{i=1}^{N_e} (e_{n|n-1}^{i})(\varepsilon_{n|n-1}^{i})^{T} \tag{6}$$

$$P_{n|n-1}^{\varepsilon,\varepsilon} = \frac{1}{N_e - 1} \sum_{i=1}^{N_e} (\varepsilon_{n|n-1}^{i})(\varepsilon_{n|n-1}^{i})^{T} \tag{7}$$

with the two covariance matrices, the Kalman gain is calculated as:

$$K = P_{n|n-1}^{e,\varepsilon}(P_{n|n-1}^{\varepsilon,\varepsilon} + R)^{-1} \tag{8}$$

where R is the covariance of the measurement noise.

Each member of the ensemble is updated as:

$$x_{n|n}^{i} = x_{n|n-1}^{i} + K(y_{n}^{obs} - \hat{y}_{n|n-1}^{i}) \tag{9}$$

where y_{n}^{obs} is the true measurement value at time step n.

2.3. Gaussian Mixture Model Based Ensemble Kalman Filter (EnKF-GMM)

2.3.1. Expectation Maximization (EM) for Clustering of the Gaussian Mixture Model

The probability distribution function of a random vector x following a finite Gaussian mixture distribution is given by:

$$p_X(x) = \sum_{j=1}^{M} \pi_j \times N\left(x; \mu_j, P_j\right) \tag{10}$$

subject to constraints that $\pi_j \geqslant 0$ and $\sum_{j=1}^{M} \pi_j = 1$, where π_j, μ_j, P_j are the prior probability, mean and covariance of mode j and $N\left(x; \mu_j, P_j\right) = \dfrac{1}{(2\pi)^{n/2} |P_j|^{1/2}} e^{-\frac{1}{2}(x-\mu_j)^T P^{-1}(x-\mu_j)}$.

Given a set of data $\{x_i\}_{i=1,...,N}$ randomly generated by a GMM, the expectation maximization (EM) algorithm is used to estimate the parameters of the GMM, $\theta = \{\pi_1, \ldots, \pi_M, \mu_1, \ldots, \mu_M, P_1, \ldots, P_M\}$ [24]. EM is a variant of maximum likelihood estimation when there exist hidden variables or missing data. In this case, the mode identity of each data point is considered as the missing or hidden variable. Let $\{(c_i)_j\}$

134

be a binary indicator vector representing the identity of the component that generates x_i. Its value is given by:

$$(c_i)_j = \begin{cases} 1, & if\ data\ point\ is\ generated\ by\ component\ j \\ 0, & otherwise \end{cases} \tag{11}$$

In the EM algorithm, an E-step is performed first to compute the Q function, the expectation of the log likelihood of the complete data set, by computing the probability of each data x_i belonging to each component j given the current parameters θ^k estimated from the previous iteration. Specifically, $Q(\theta|\theta^k) = E[L(p(z|\theta))|\{x\}, \theta^k]$, where $\{x\}$ is the observed data set; $\{z\}$ is the complete data set consisting of both observed and missing data, $\{z\} = \{c_1, x_1, \ldots, c_N, x_N\}$, c_i is the membership of each data point, and θ^k is the estimate of the last iteration. This becomes

$$Q\left(\theta|\theta^k\right) = \sum_{i=1}^{N} \sum_{j=1}^{M} p[(c_i)_j|\{x\},\ \theta^k]\left(log\pi_j N\left(x_i; \mu_j, P_j\right)\right) \tag{12}$$

$$w_{ij} = p[(c_i)_j|\{x\},\ \theta^k] = \frac{\pi_j^k N\left(x_i; \mu_j^k, P_j^k\right)}{\sum_{m=1}^{M} \pi_m^k N\left(x_i; \mu_m^k, P_m^k\right)} \tag{13}$$

Next, the M-step is performed to maximize the Q function and calculate the corresponding θ^{k+1}.

$$\pi_j^{k+1} = \frac{N^k}{N} \tag{14}$$

$$\mu_j^{k+1} = \frac{1}{N^k} \sum_{i=1}^{N} w_{ij} x_i \tag{15}$$

$$P_j^{k+1} = \frac{1}{N^k} \sum_{i=1}^{N} w_{ij}(x_i - \mu_j^{k+1})(x_i - \mu_j^{k+1})^T \tag{16}$$

where $N^k = \sum_{i=1}^{N} w_{ij}$.

The E-step and the M-step are performed iteratively until the estimates converge. During this process, the problem of singularity may arise when one of the components collapses onto one data point. This usually happens due to over-fitting in the maximum likelihood estimation (MLE). To avoid this problem, one approach is to adopt a Bayesian regularization method [25] to replace the MLE with the maximum *a posteriori* (MAP) estimate. Based on this method, the update of the covariance is modified to become

$$P_j^{k+1} = \frac{\sum_{i=1}^{N} w_{ij}(x_i - \mu_j^{k+1})(x_i - \mu_j^{k+1})^T + \lambda I_d}{N^k + 1} \tag{17}$$

where I_d is an n-dimensional unit matrix and λ is a regularization constant determined by some validation data [26]. An alternate (*ad hoc*) method to deal with the problem of singularity is to detect when the singularity occurs and reset the means of all components randomly and the covariance to some larger value.

The pseudo-code for the EM algorithm is provided below.

Algorithm 1: Expectation Maximization algorithm. Inputs are data set $\{x_i\}_{i=1,..,N}$, component number M and initial values $\{\theta^0\}$ of $\{\pi_j\}_{j=1,...,M}$, $\{\mu_j\}_{j=1,...M}$, $\{P_j\}_{j=1,...\ M}$, $\theta^k = \theta^0$.

EM$[\{x\}, M, \{\theta^k\}]$
// *E* step
while $\varepsilon \leqslant 1e - 6$
for $i = 1: N$
    for $j = 1:M$
        $p[(c_i)_j | x_i, \theta^k] = p(x_i | (c_i)_j, \theta^k) p((c_i)_j | \theta^k) / p(x_i)$
    end for
end for
// M step
for $j = 1:M$

$$\pi_j^{k+1} = \sum_{i=1}^{N} p[(c_i)_j | x_i, \theta^k]/N$$

$$\mu_j^{k+1} = \sum_{i=1}^{N} p[(c_i)_j | x_i, \theta^k] x_i / \sum_{i=1}^{N} p[(c_i)_j | x_i, \theta^k]$$

$$P_j^{k+1} = \frac{\sum_{i=1}^{N} p[(c_i)_j | x_i, \theta^k] \left(x_i - \mu_j^{k+1}\right)\left(x_i - \mu_j^{k+1}\right)^T + \lambda I_d}{\sum_{i=1}^{N} p[(c_i)_j | x_i, \theta^k] + 1}$$

end for

$$\varepsilon = \mu^{k+1} - \mu^k$$

end while
return θ^{k+1}

2.3.2. EnKF-GMM Algorithm

In this section, a GMM-based EnKF (EnKF-GMM) filter is proposed to obtain estimates of the full state distribution. As with the particle filter, it also uses a set of particles to represent the posterior probability distribution function (PDF) of the states. The difference is that the PDF is constrained to be a GMM at every time step.

At each time step, the EnKF-GMM has two steps—forecast and update. The forecast step is identical to the EnKF. An ensemble of size N, $\{x_i\}_{i=1,...,N}$, is drawn from the prior distribution of the states and forwarded through the model to obtain a predicted ensemble for the next time step. Then, the EM algorithm is performed on the predicted ensemble to obtain the estimates of the GMM with M components.

Next, the Kalman update is performed based on each component in the GMM to get an ensemble of size $N \times M$. Finally, these ensemble members are combined based on their weights and reduced to a size of N. The details of the algorithmic sequence are as follows:

Forecast:

1. The first portion of the forecast step is to determine the number of components M in the multimodal distribution. M can be determined using the Bayesian or other information criteria [27,28], or using prior knowledge. For example, in reservoir models, petrophysical properties (such as porosity or permeability) are typically related to geological units (facies), and variables inside the facies are characterized by underlying multimodal distributions which are known beforehand [9]. In our work, this information can be considered as prior knowledge if we know the distribution of the process noise.
2. With the knowledge of the process model and the number of components M, the prior ensemble $\{x_i\}_{i=1,...,N}$ is propagated through the model to get the predicted values of the ensemble $\{x_i^f\}_{i=1,...,N}$. These are the realizations of the predicted state space x^f.

Assuming the predicted state x^f at the forecast step is a GMM,

$$p\left(x^f\right) = \sum_{j=1}^{M} \tau_j^f p_j\left(x^f\right) = \sum_{j=1}^{M} \tau_j N(x^f; \mu_j^f, P_j^f) \tag{18}$$

The EM algorithm is applied on $\{x_i^f\}_{i=1,...,N}$ to give us the parameters of the prior distribution (τ_j^f, μ_j^f and P_j^f) of each component j.

Update:

3. For each component j of the distribution, the Kalman gain matrix for each Gaussian component is computed by utilizing the membership probability matrix W.

$$P[j]^f H^T = \sum_{i=1}^{N} w_{i,j}(x_i^f - \mu_j)(Hx_i^f - H\mu_j)^T / n_j \tag{19}$$

$$HP[j]^f H^T = \sum_{i=1}^{N} w_{i,j}(Hx_i^f - H\mu_j)(Hx_i^f - H\mu_j)^T / n_j \tag{20}$$

$$K[j] = P[j]^f H^T (HP[j]^f H^T + R)^{-1} \tag{21}$$

where $w_{i,j} = \dfrac{\pi_j N\left(x_i; \mu_j, P_j\right)}{\sum_{m=1}^{M} \pi_m N\left(x_i; \mu_m, P_m\right)}$, $n_j = \sum_{i=1}^{N} w_{i,j}$, and H is the linearized measurement function.

4. In the update step, assuming one Gaussian component j claims the ownership of all the ensemble members, the Kalman update can be performed for each component member under component j. This gives us an ensemble size of $N \times M$.

$$x_i^{a,j} = x_i^f + K[k](d - Hx_i^f - e_i) \tag{22}$$

5. The $N \times M$ ensemble members can be combined to form N members by using the probability matrix. This gives us the final posterior ensemble $\{x_i^a\}_{i=1,...,N}$.

$$x_i^a = \sum_{j=1}^{M} w_{i,j} x_i^{a,j} \tag{23}$$

The mean and covariance of the posterior can be computed as:

$$\mu_j^a = \sum_{i=1}^{N} w_{i,j} x_i^{a,j} / n_j \tag{24}$$

$$P[j]^a = \sum_{i=1}^{N} w_{i,j} (x_i^{a,j} - \mu_j^a)(x_i^{a,j} - \mu_j^a)^T / n_j \tag{25}$$

6. The posterior weight of each component of the distribution can be computed based on the observed data d, which contains the measurements y.

$$\tau_j^a = p\left(\mu_j, \Sigma_j, R|d\right) = \frac{p(d|\mu_j, \Sigma_j, R)n_j}{\sum_{j=1}^{M} p(d|\mu_j, \Sigma_j, R)n_j} \tag{26}$$

$$p(d|\mu_j, \Sigma_j, R) = \frac{exp[-\frac{1}{2}\left(d - H\mu_j\right)^T \left(H\Sigma_j H^T + R\right)^{-1}\left(d - H\mu_j\right)]}{\sqrt{(2\pi)^m |H\Sigma_j H^T + R|}} \tag{27}$$

7. The point estimate is given by:

$$x^a = \sum_{j=1}^{M} \tau_j^a \mu_j^a \tag{28}$$

The pseudo-code for the EnKF-GMM algorithm is provided below.

Algorithm 2: EnKF-GMM algorithm. Inputs include the initial distribution of x, the total number of the particles N, the components M, and the time steps T. Inputs and observations at each time step are u_n and d_n.

$$[\{x_i^a\}_{i=1}^N, \{\mu_j^a, P_j^a . \tau_j^a\}_{j=1}^M] = \textbf{EnKF-GMM}[\{x_i\}_{i=1}^N, d_t]$$

for n = 1:T

    for I = 1 : N

        Draw $x_i^f \sim f\left(I, u_{n-1}, v_{n-1}^i\right)$

        Calculate $y_i = Hx_i^f + v_n^i$

    end for

    Apply the EM algorithm on $\{x_i^f\}_{i=1,\dots,N}$ using algorithm 1:

    $\{\tau_j^f, \mu_j^f, P_j^f\}_{j=1}^M = EM[\{x_i^f\}_{i=1,\dots,N}, M, \{\theta^k\}]$

    for j = 1 : M

        Calculate the Kalman gain of each component $K[j]$ using Equation (21)

        foI i = 1 : N

            Calculate the updated particles for each component $\{x_i^{a,j}\}_{i=1}^N$ using Equation (22)

        end for

        Combine $\{x_i^{a,j}\}_{i=1}^N$ to obtain the posterior particles $\{x_i^a\}_{i=1}^N$ using Equation (23)

        Calculate the parameters of the posterior distribution $\mu_j^a, P_j^a . \tau_j^a$ using Equations (24)–(26).

    end for

    Calculate the point estimate x^a using Equation (28)

    end for

While the PF and the EnKF-GMM both can, in principle, account for multimodality, the use of the Gaussian mixture model provides the EnKF-GMM with greater flexibility in capturing a wide variety of distributions under varying levels of model-plant mismatch, as will be shown in the results.

3. Results and Discussion

3.1. Mathematical Model of the Methyl Methacrylate (MMA) Polymerization Process

Simulations of a free-radical methyl methacrylate (MMA) polymerization process are used to demonstrate the performance of the estimation method proposed in this paper. The process is assumed to take place in a continuous stirred tank reactor (CSTR), and uses AIBN as the initiator and toluene as the solvent. The mathematical model of this process is described below in Equations (29)–(35), and further details can be found in [29,30]. Parameter values are provided in Table 1. The six states to be estimated include the monomer concentration C_M, the initiator concentration C_I, the reactor temperature T, the moments of the polymer distribution, D_0 and D_1, and the jacket temperature T_j. Only the temperatures are measured. The number average molecular weight (NAMW), which is the primary quality variable for the process, is defined as the ratio D_1/D_0.

$$\frac{dC_m}{dt} = -\left(k_p - k_{fm}\right) C_m P_0 + \frac{F\left(C_{min} - C_m\right)}{V} \tag{29}$$

$$\frac{dC_I}{dt} = -k_I C_I + \frac{\left(F_I C_{Iin} - F C_I\right)}{V} \tag{30}$$

$$\frac{dT}{dt} = \frac{(-\Delta H)\, k_p C_m P_0}{\rho C_\rho} - \frac{UA}{\rho C_\rho V}\left(T - T_j\right) + \frac{F\left(T_{in} - T\right)}{V} \tag{31}$$

$$\frac{dD_0}{dt} = \left(0.5 k_{tc} + k_{td}\right) P_0^2 + k_{fm} C_m P_0 - \frac{F D_0}{V} \tag{32}$$

$$\frac{dD_1}{dt} = M_{rt}\left(k_p + k_{fm}\right) C_m P_0 - \frac{F D_f}{V} \tag{33}$$

$$\frac{dT_j}{dt} = \frac{F_{cw}\left(T_{w0} - T_j\right)}{V_0} + \frac{UA}{\rho_w C_{pw} V_0}\left(T - T_j\right) \tag{34}$$

$$P_0 = \sqrt{\frac{2f^* + C_I k_I}{k_{td} + k_{tc}}} \tag{35}$$

Table 1. Operational parameters for the methyl methacrylate (MMA) polymerization reactor.

$F = 1.0\ m^3/h$	$M_m = 100.12\ kg/kgmol$
$F_I = 0.0032\ m^3/h$	$f^* = 0.58$
$F_{cw} = 0.1588\ m^3/h$	$R = 8.314\ kJ/kgmol \cdot K$
$C_{min} = 6.4678\ kgmol/m^3$	$-\Delta H = 57800\ kJ/kgmol$
$C_{Iin} = 8.0\ kgmol/m^3$	$E_p = 1.8283 \times 10^4\ kJ/kgmol$
$T_{in} = 350\ K$	$E_I = 1.2877 \times 10^5\ kJ/kgmol$
$T_{w0} = 293.2\ K$	$E_{fm} = 7.4478 \times 10^4\ kJ/kgmol$
$U = 720\ kJ/h \cdot K \cdot m^2$	$E_{tc} = 2.9442 \times 10^3\ kJ/kgmol$
$A = 2.0\ m^2$	$E_{td} = 2.9442 \times 10^3\ kJ/kgmol$
$V = 0.1\ m^3$	$A_p = 1.77 \times 10^9\ m^3/kgmol \cdot h$
$V_0 = 0.02\ m^3$	$A_I = 3.792 \times 10^{18}\ 1/h$
$\rho = 866\ kg/m^3$	$A_{fm} = 1.0067 \times 10^{15}\ m^3/kgmol \cdot h$
$\rho_w = 1000\ kg/m^3$	$A_{tc} = 3.8223 \times 10^{10}\ m^3/kgmol \cdot h$
$C_p = 2.0\ kJ/(kg \cdot K)$	$A_{td} = 3.1457 \times 10^{11}\ m^3/kgmol \cdot h$
$C_{pw} = 4.2\ kJ/(kg \cdot K)$	

In all the simulations whose results are described in the following sections, the number of particles used for each estimator, N, is 100. The number of components, M, is set to 2. The parameters of the bimodal noise in all simulations are $\mu=[0.1,0.8]$, $P=\text{diag}(0.1,0.1)$ for states C_m, C_I and D_0; $\mu=[8,64]$, $P=\text{diag}(8,8)$ for state D_1; and $\mu=[0.6,4.8]$, $P=\text{diag}(0.6,0.6)$ for states T and Tj.

The simulations we perform are introduced here: Case Study 1 provides a comparison of the EnKF-GMM, the PF, and the EnKF for a case with bimodal distributions and insignificant model-plant mismatch. Case Study 2 provides a comparison of the three estimators where the model-plant mismatch is significant. Case Study 3 compares the estimators for state estimation with uncertain parameters, but with the uncertain parameter not being estimated. Case Study 4 considers the same case as Case Study 3, but with combined state and parameter estimation. In Case Study 5, we consider an alternate version of the PF and use the simulation conditions of Case Study 2.

3.2. Comparison of State Estimation with the EnKF-GMM, EnKF, and PF (Case Study 1)

In this section, we present the results of applying the EnKF-GMM, EnKF, and PF algorithms on the PMMA process. To illustrate the performance of the estimators in cases where the states have multimodal distributions, bimodal process noise is applied to all the six states. The measurement noise is assumed to be Gaussian. The prior distribution of the state is also assumed to follow a GM distribution which contains two modes.

For Case Study 1, the true initial values of the states are:

$$x_0 = [5\frac{\text{kgmol}}{m^3}, \ 3\frac{\text{kgmol}}{m^3}, \ 320\text{K}, \ 0.5\ \frac{\text{kgmol}}{m^3}, \ 0.5\frac{\text{kg}}{m^3}, \ 300 \text{ K}]$$

The dynamics of the simulation describe how the system relaxes to a steady state from this initial condition. For the estimators, the initial particles are drawn from the prior distribution. The tuning parameters for the prior distribution are its mean and covariance. In the first case, a prior distribution with a small amount of bimodal process noise is tested for the three algorithms. The means of the two Gaussian modes of the prior distribution are:

$$\mu_1 = \left[4\ \frac{\text{kgmol}}{m^3}, \ 2\ \frac{\text{kgmol}}{m^3}, \ 310 \text{ K}, \ 0.49\ \frac{\text{kgmol}}{m^3}, \ 0.49\ \frac{\text{kg}}{m^3}, \ 295 \text{ K}\right];$$

$$\mu_2 = \left[6\ \frac{\text{kgmol}}{m^3}, \ 4\ \frac{\text{kgmol}}{m^3}, \ 330 \text{ K}, \ 0.51\ \frac{\text{kgmol}}{m^3}, \ 0.51\ \frac{\text{kg}}{m^3}, \ 305 \text{ K}\right]$$

The covariances of the modes of the prior distribution are:

$$P_1 = \mathrm{diag}\,(4,\ 4,\ 28,\ 8e-1, 8e-4,\ 6)\,;$$

$$P_2 = \mathrm{diag}\,(4,\ 4,\ 28,\ 8e-1, 8e-4,\ 6)$$

The tuning parameters of the initial distribution indicate a state distribution with insignificant bimodality. The purpose of this simulation is to demonstrate the estimation performance of the three algorithms in the scenario where the state distribution shows insignificant multimodality.

The comparison of estimation results using the EnKF-GMM, EnKF, and PF is shown in Figure 1, with time steps on the x-axis (each time step is 0.3 h = 18 min). Table 2 shows the root mean squared error (RMSE) over the 25 time steps of the simulation for the six states and the NAMW for the three algorithms. In this case, the estimation results from Figure 1 and Table 2 show that the three algorithms have similar performance in estimation of the six states. However, the EnKF-GMM has the best performance in the estimation of the NAMW. In addition, the converged variance of the estimates of the states, obtained from the estimated covariance matrix with the EnKF-GMM, are $[10^{-4}, 10^{-4}, 1.2 \times 10^{-4}, 10^{-5}, 2 \times 10^{-4}, 4 \times 10^{-4}]$, respectively, confirming the significance of the estimates. The PF performs better than the EnKF only for some states. Increasing the number of particles for each of the algorithms to 200 (results not shown) improves the performance of the PF slightly, but the same conclusions hold.

Table 2. RMSE of the Gaussian mixture model based ensemble Kalman filter (EnKF-GMM), ensemble Kalman filter (EnKF), and particle filter (PF) for the polymethyl methacrylate (PMMA) process with multimodal process noise (Case Study 1).

Variable	EnKF-GMM	EnKF	PF
C_M, $kg \cdot mol/m^3$	0.20	0.20	0.33
C_I, $kg \cdot mol/m^3$	0.24	0.20	0.33
T, K	4.3	4.4	3.1
D_0, $kg \cdot mol/m^3$	0.019	0.014	0.032
D_1, kg/m^3	11.85	11.53	10.44
T_j, K	2.3	2.2	1.4
NAMW	209	338	357

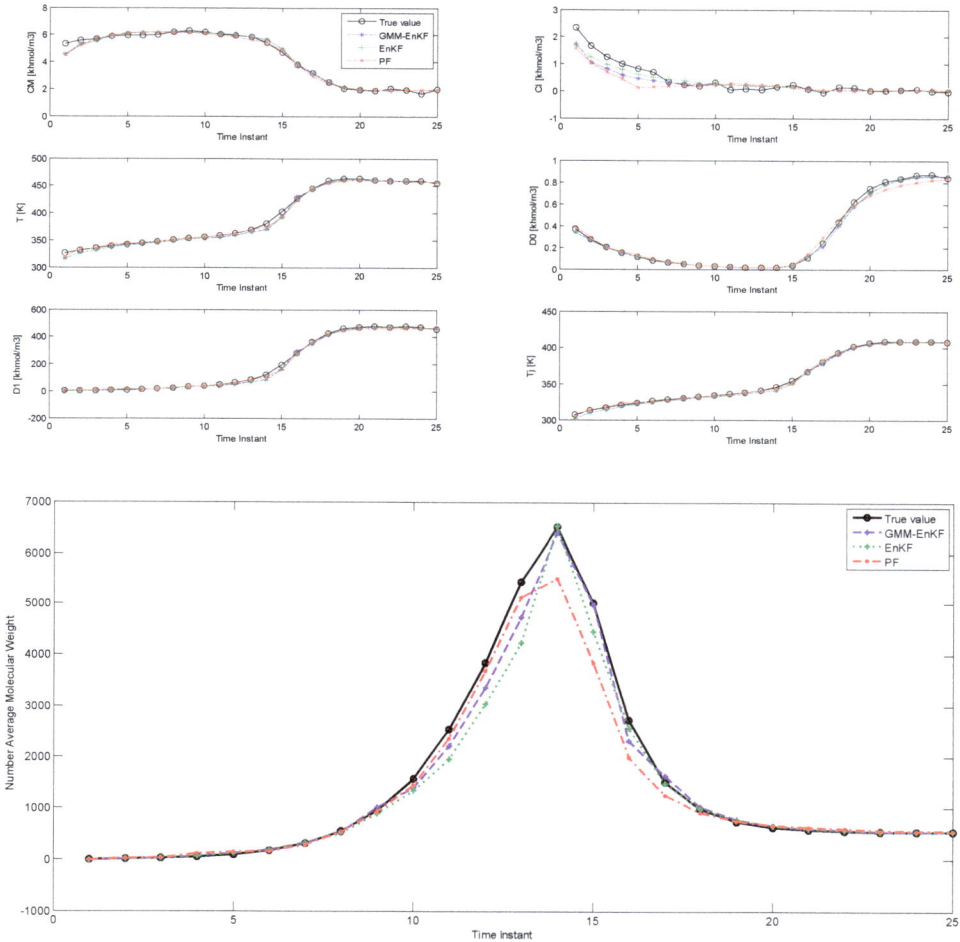

Figure 1. Comparison of the estimation performance of the ensemble Kalman filter (EnKF)-Gaussian mixture model (GMM), EnKF, and particle filter (PF) for the polymethyl methacrylate (PMMA) process with multimodal process noise (Case Study 1).

In Case Study 2, the multimodal features of the prior distribution are made more significant compared with the first case. The parameters of the prior distribution given below indicate that both modes lie far away from the true value, which also means that the initial condition mismatch is much larger. The true initial values of the states remain the same as the first case, and the process noise and measurement noise applied to the plant remain unchanged as well. The modified prior distribution is specified by:

143

$$\mu_1 = \left[1\ \frac{kgmol}{m^3},\ 1\ \frac{kgmol}{m^3},\ 290\ \text{K},\ 0.49\ \frac{kgmol}{m^3},\ 0.49\ \frac{kg}{m^3},\ 270\ \text{K}\right];$$

$$\mu_2 = \left[10\ \frac{kgmol}{m^3},\ 8\ \frac{kgmol}{m^3},\ 350\ \text{K},\ 0.51\ \frac{kgmol}{m^3},\ 0.51\ \frac{kg}{m^3},\ 330\ \text{K}\right];$$

$$P_1 = \text{diag}\,(0.8,\ 0.8,\ 5.6,\ 8e-2, 8e-3,\ 5.6)\,;$$

$$P_2 = \text{diag}\,(0.8,\ 0.8,\ 5.6,\ 8e-2, 8e-3,\ 5.6)$$

In this case, the parameters of the prior distribution indicate that both of the modes lie near the tail of the likelihood function. The initial particles not only show significant multimodality, but also some degree of model-plant mismatch. The comparison of estimation using the EnKF-GMM, EnKF, and PF is shown in Figure 2 and the RMSE is shown in Table 3, and it is clear that the EnKF-GMM outperforms the other two estimators. As expected, the performance of the EnKF has worsened in this case because its Gaussian assumption on the prior and posterior distributions is violated in a significant manner. The PF does not show good performance either, and it is outperformed by the EnKF in the estimation of the NAMW. This is because the PF lacks robustness to plant-model mismatch [14], which is present in this case. Increasing the number of particles for all the estimators does not change these conclusions.

Figure 3 shows the evolution of the multimodal posterior distribution of the one of the states (the monomer concentration) at time steps 1, 3, 4, and 9. Table 4 lists the corresponding estimation errors of the three algorithms at those time steps with respect to the true value of C_M. Figure 4 shows the evolution of the posterior distribution of another state (the jacket temperature) at time steps 2, 6, 9, and 10, and Table 5 shows the corresponding estimation errors of the three algorithms. These distributions are bimodal, and this clearly shows that the EnKF-GMM outperforms the other estimators in the presence of multimodal distributions.

Table 3. RMSE of the EnKF-GMM, EnKF, and PF for the PMMA process with more significant multimodal process noise (Case Study 2).

Variable	EnKF-GMM	EnKF	PF
$C_M,\ kg\cdot mol/m^3$	0.44	0.68	0.69
$C_I,\ kg\cdot mol/m^3$	0.37	0.14	0.17
$T,\ K$	5.8	11.8	14.4
$D_0,\ kg\cdot mol/m^3$	0.042	0.062	0.078
$D_1,\ kg/m^3$	9.73	36.13	51.38
$T_j,\ K$	5.1	8.2	9.2
NAMW	559	1400	831

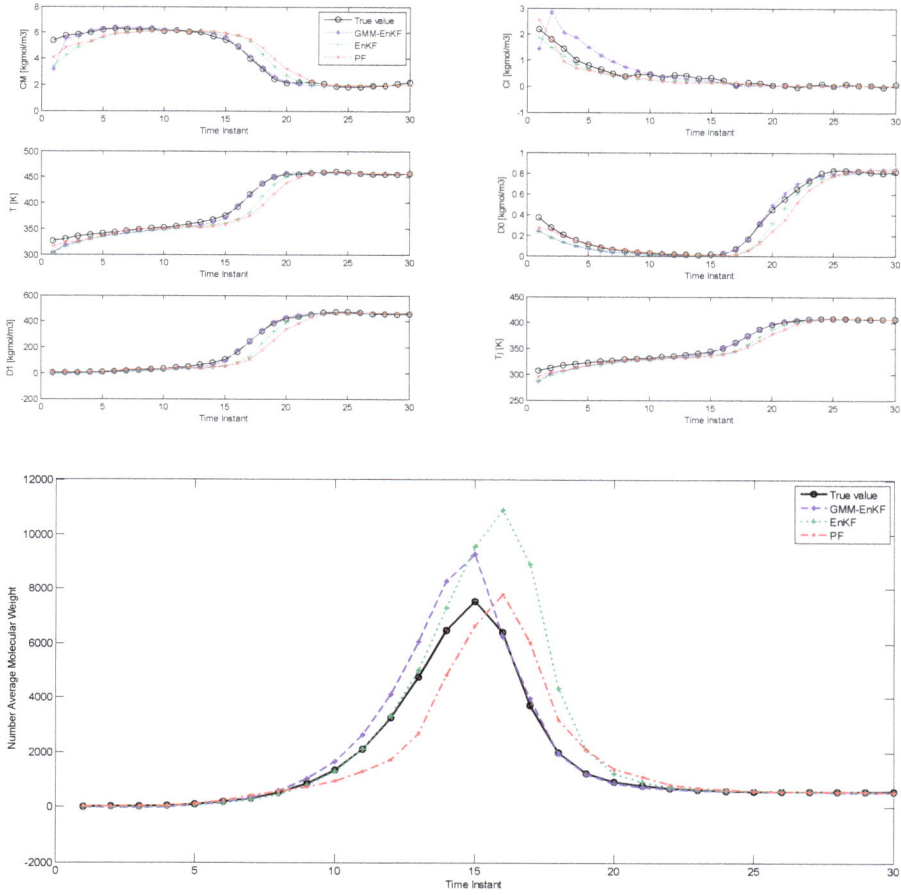

Figure 2. Comparison of the estimation performance of the EnKF-GMM, EnKF, and PF for the PMMA process with more significant multimodal process noise (Case Study 2).

Table 4. Comparison of the estimation errors of the EnKF-GMM, EnKF, and PF for C_M at time steps 1, 3, 4, and 9 (in kg·mol/m^3) (Case Study 2).

Estimator	Time Step 1	Time Step 3	Time Step 4	Time Step 9
EnKF-GMM	0.23	0.14	0.40	0.04
EnKF	2.06	1.06	0.80	0.10
PF	3.60	2.20	1.65	0.22

145

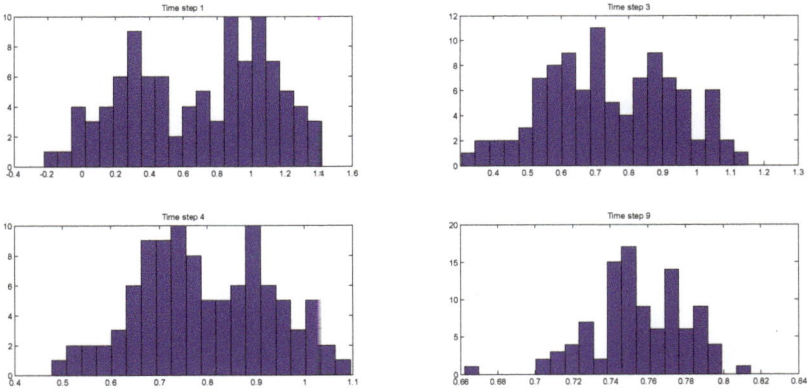

Figure 3. Evolution of the multimodal posterior distributions of C_M at time steps 1, 2, 4, and 9 (Case Study 2).

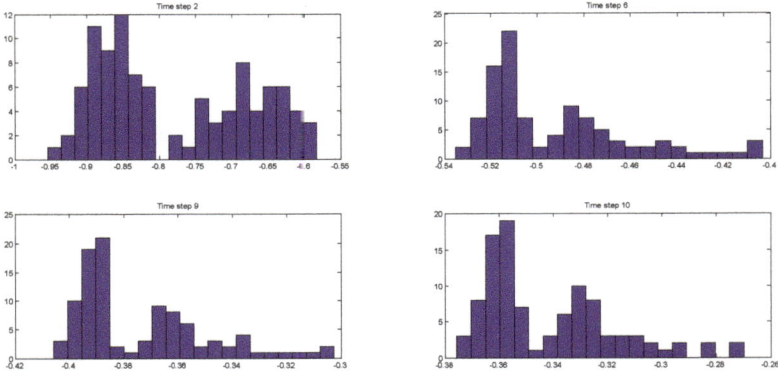

Figure 4. Evolution of the multimodal posterior distributions of T_j at time steps 2, 6, 9, and 10 (Case Study 2).

Table 5. Comparison of the estimation errors of the EnKF-GMM, EnKF, and PF for T_j at time steps 2, 6, 9, and 10 (ir. K) (Case Study 2).

Estimator	Time Step 1	Time Step 3	Time Step 4	Time Step 9
EnKF-GMM	6.4	2.8	1.5	1.3
EnKF	6.6	3.0	1.9	1.7
PF	13.5	4.5	2.9	2.9

146

3.3. Comparison of State and Parameter Estimation with the EnKF-GMM, EnKF and PF (Case Studies 3 and 4)

We consider the effects of parametric uncertainty in this section. The uncertain parameter chosen for these studies is E_p, which is the activation energy associated with the reaction rate parameter k_p. We choose E_p as the uncertain parameter because (based on dimensionless sensitivity analysis) the NAMW is highly sensitive to the values of this parameter. We consider state estimation and joint state and parameter estimation in this section.

3.3.1. State Estimation with Uncertain Parameter (Case Study 3)

In this sub-section, while E_p is an uncertain parameter and noise is added to its value at each time step in the simulation, the parameter is not estimated. The nominal value of E_p is set to be $E_p = 1.8283 \times \dfrac{10^4 \text{kJ}}{\text{kgmol}}$, and bimodal Gaussian noise with means of the modes $\mu_1 = -100$, $\mu_2 = 100$ and covariances $P_1 = 50$, $P_2 = 50$ is added to it. In addition, process and measurement noise with the same distributions as in the second case in Section 3.2 are included. Figure 5 shows the comparison of the estimation results using the three algorithms over 40 time steps, and Table 6 shows the corresponding RMSE. In this case, the EnKF-GMM shows a small improvement in state estimation performance over the other estimators, especially in the estimation of the NAMW.

Table 6. RMSE of the EnKF-GMM, EnKF, and PF for state estimation in the case with uncertain parameter E_p (Case Study 3).

Variable	EnKF-GMM	EnKF	PF
C_M, $kg \cdot mol/m^3$	0.29	0.26	0.32
C_I, $kg \cdot mol/m^3$	0.12	0.10	0.27
T, K	7.2	8.9	10.3
D_0, $kg \cdot mol/m^3$	0.111	0.092	0.144
D_1, kg/m^3	32.27	35.11	45.34
T_j, K	5.5	5.7	7.5
NAMW	487	869	653

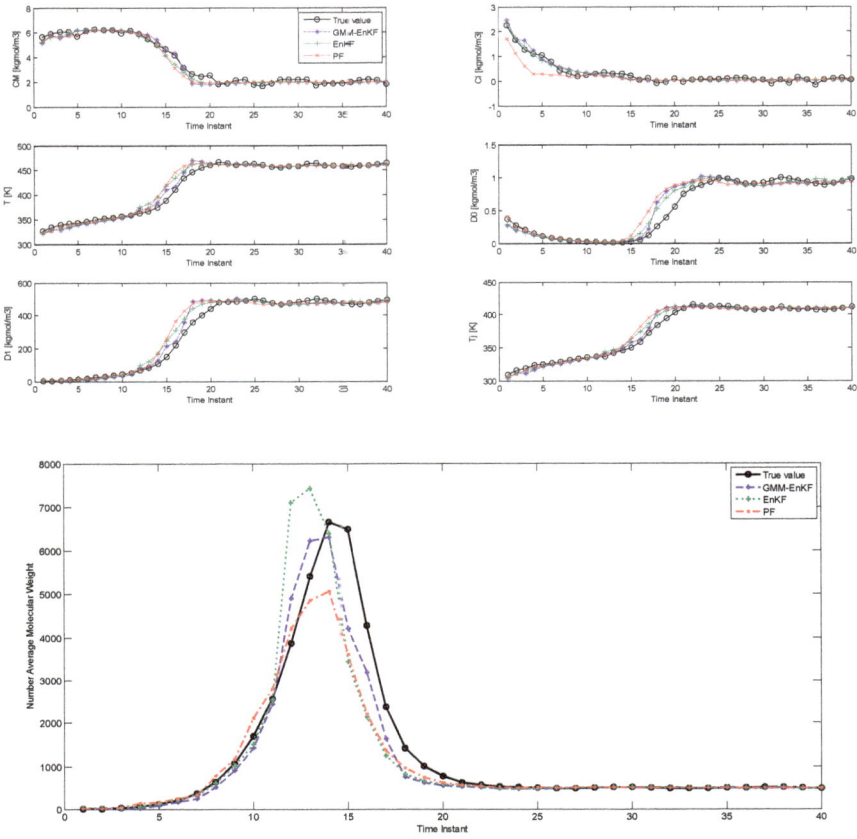

Figure 5. Comparison of state estimation with the EnKF-GMM, EnKF, and PF for the PMMA process with uncertain parameter E_p (Case Study 3).

3.3.2. State and Parameter Estimation with Uncertain Parameter (Case Study 4)

Next, we compare the performance of the estimators for joint state and parameter estimation. Once again, E_p is the uncertain parameter and its nominal value is kept the same as in Case Study 3. The parameter E_p is treated as an augmented state for estimation. The prior distribution for E_p has the following characteristics: means of $\mu_1 = 1.9 \times 10^4$, $\mu_2 = 2.5 \times 10^4$ for its two modes, and covariances of $P_1 = 500$, $P_2 = 500$. Bimodal noise is added to each particle of the parameter, with means $\mu_1 = -100$, $\mu_2 = 100$ and covariances $P_1 = 50$, $P_2 = 50$. Except for the exclusion of process noise, the properties of the simulation are kept the same as in Case Study 3. Figure 6 shows the performance of the estimators in state estimation, and Figure 7 shows their performance in estimating the parameter E_p. While the performance of the EnKF in state estimation is comparable to that of

148

the EnKF-GMM, the EnKF-GMM is clearly superior in parameter estimation. The PF has the worst performance among the estimators.

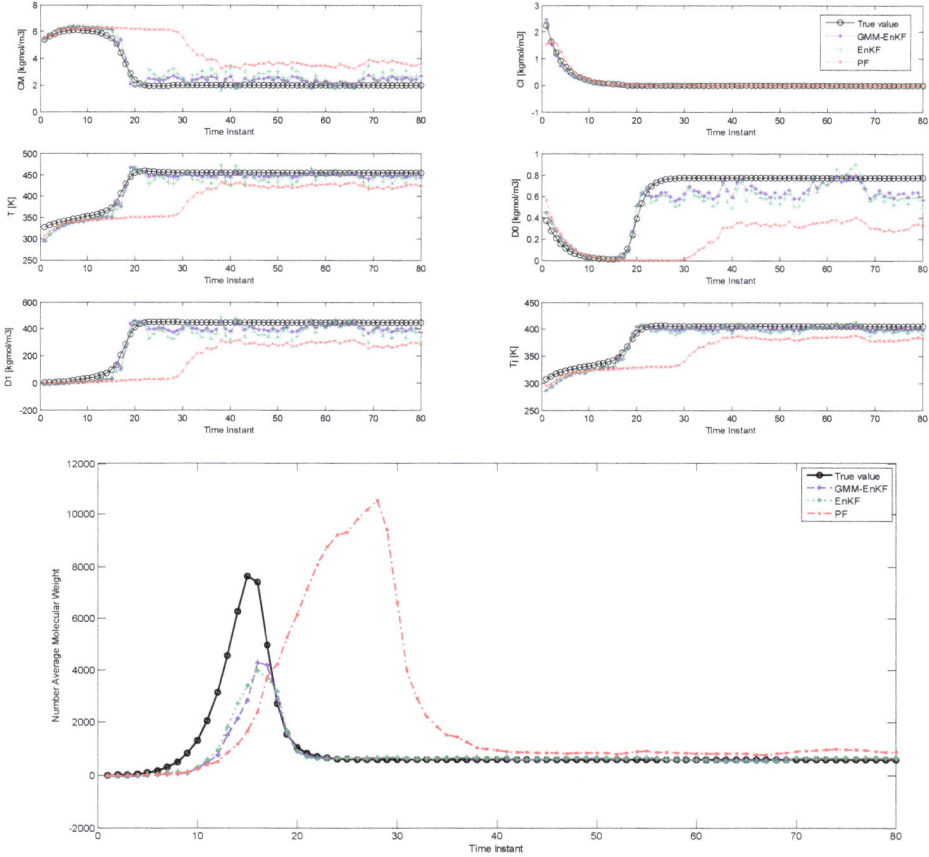

Figure 6. Comparison of state estimation with the EnKF-GMM, EnKF, and PF for the PMMA process with uncertain parameters (Case Study 4).

3.4. Alternate Point Estimates for the PF (Case Study 5)

In the PF, even though the full distribution is obtained, a point estimate for the states is usually obtained by choosing the expectation (mean) of the posterior particles. This is the method we have employed for the PF in the simulations described in the previous sections. However, if the distribution is multimodal, the mean may not necessarily represent the best point estimate, and the mode of the distribution (which is equivalent to the maximum *a posteriori* estimate) can provide a better estimate [14,31]. We investigate whether this approach can improve the performance of the PF, since we are considering cases where the distributions are multimodal. We

149

apply *k*-means clustering on the posterior distribution of the particles to identify the modes and the maximum *a posteriori* estimate with the particle filter, and compare the estimation performance of this PF, called the PF-mode, with the other estimators. The parameters of the simulations are similar to the second case study. Figure 8 shows the performance of the estimators, and the RMSE is described in Table 7. The PF-mode clearly outperforms the PF and the EnKF; however, the EnKF-GMM has superior performance.

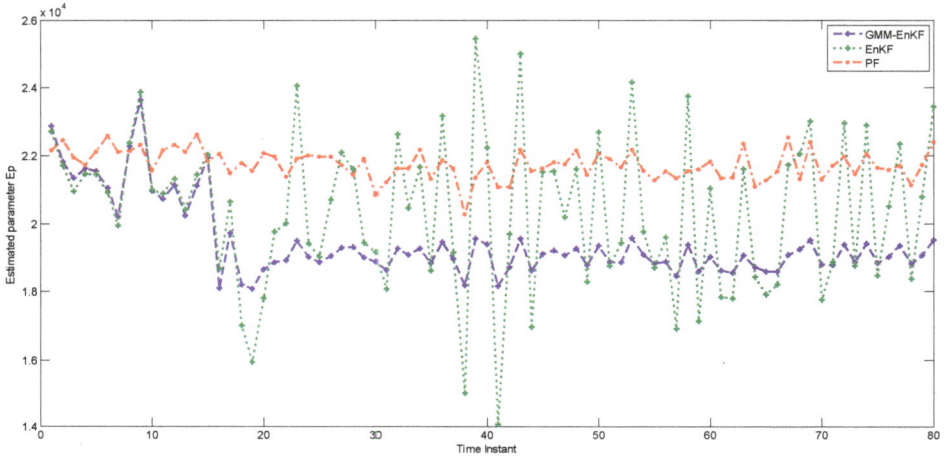

Figure 7. Parameter estimation using the EnKF-GMM, EnKF, and PF (Case Study 4).

Table 7. RMSE of the EnKF-GMM, EnKF, PF, and PF-mode for state estimation (Case Study 5).

Variable	EnKF-GMM	EnKF	PF	PF-mode
C_M, kg· mol/m^3	0.44	0.68	0.68	0.85
C_I, kg· mol/m^3	0.37	0.14	0.17	0.55
T, K	5.8	11.8	14.4	8.31
D_0, kg· mol/m^3	0.042	0.062	0.078	0.047
D_1, kg/m^3	9.73	36.13	51.38	13.05
T_j, K	5.1	8.2	9.2	7.9
NAMW	559	1400	831	706

150

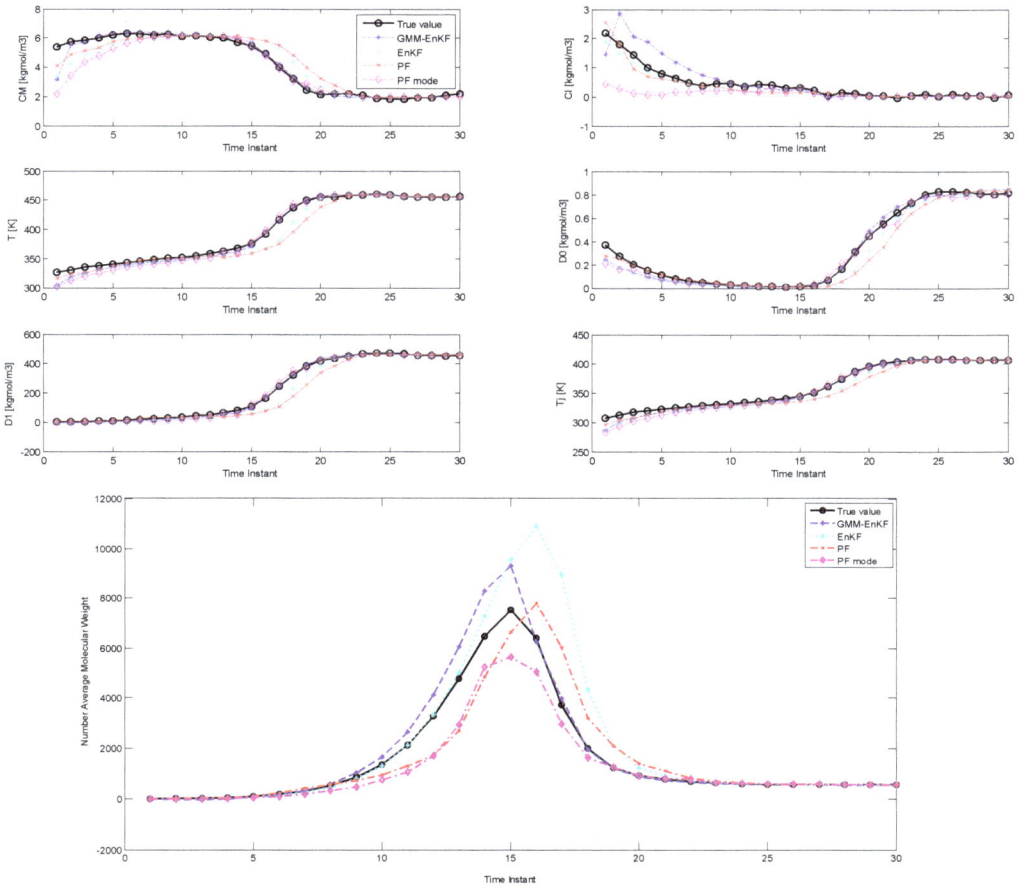

Figure 8. Comparison of state estimation with the EnKF-GMM, EnKF, PF, and PF-mode (Case Study 5).

The idea of the PF-mode is very similar to that of the EnKF-GMM. Both of them use clustering to extract modes from the posterior distribution and generate a point estimate based on the information in the modes. However, the EnKF-GMM outperforms the PF-mode because it is more robust to poor initial estimates and model-plant mismatch. Also, if the number of modes in the state distributions varies with time, perhaps even becoming unimodal at some times, using the mode as a point estimate is not necessarily superior to the mean. The EnKF-GMM combines the modes of the distribution in proportion based on the calculated weights to get a point estimate, and can adjust its estimation results in these cases by adjusting the weights of the modes.

151

4. Conclusions

We have proposed an estimator based on a Gaussian mixture model coupled with an'ensemble Kalman filter (EnKF-GMM) that is capable of handling multimodal state distributions, and demonstrated its performance in simulations on a polymethyl methacrylate process. The EnKF-GMM clearly outperforms the particle filter (PF) and the EnKF in both state and parameter estimation with multimodal distributions. The EnKF is limited by the assumption of Gaussian distributions, and the particle filter's performance is affected by its lack of robustness with respect to model-plant mismatch. A different choice for obtaining a point estimate with the particle filter, leading to a maximum *a posteriori* estimate, improves the performance of the PF, but the EnKF-GMM is still superior, indicating that it is the estimator of choice for systems with multimodal state distributions such as polymer processes.

Acknowledgments: The authors acknowledge financial support from the China Scholarship Council and the Natural Sciences and Engineering Research Council of Canada.

Author Contributions: All three authors conceived the work and participated in defining its scope. Ruoxia Li developed the algorithms and conducted the simulations described in the manuscript with inputs from Vinay Prasad and Biao Huang. Ruoxia Li wrote the initial drafts of the manuscript, and all authors contributed to the editing of the final manuscript and to the revisions.

References

1. Wilson, D.; Agarwal, M.; Rippin, D. Experiences implementing the extended Kalman filter on an industrial batch reactor. *Comput. Chem. Eng.* **1998**, *22*, 1653–1672.
2. Prasad, V.; Schley, M.; Russo, L.P.; Bequette, W.B. Product property and production rate control of styrene polymerization. *J. Process Control* **2002**, *12*, 353–372.
3. Jo, J.; Bankoff, S. Digital monitoring and estimation of polymerization reactor. *AIChE J.* **1976**, *22*, 361–368.
4. Kozub, D.; MacGregor, J. State estimation for semi-batch polymerization reactors. *Chem. Eng. Sci.* **1992**, *47*, 1047–1062.
5. McAuley, K.; MacGregor, J. On-line inference of polymer properties in an industrial polymer properties in an industrial polyethylene reactor. *AIChE J.* **1991**, *37*, 825–835.
6. McAuley, K.; MacGregor, J. Nonlinear product property control in industrial gas-phase polyethylene reactors. *AIChE J.* **1993**, *39*, 855–866.
7. Sriniwas, G.; Arkun, Y.; Schork, F. Estimation and control of an alpha-olefin polymerization reactor. *J. Process Control* **1994**, *5*, 303–313.
8. Gopalakrishnan, A.; Kaisare, N.S.; Narasimhan, S. Incorporating delayed and infrequent measurements in extended Kalman filter based nonlinear state estimation. *J. Process Control* **2011**, *21*, 119–129.

9. Evensen, G. Sequential data assimilation with a nonlinear quasi-geostrophic model using Monte Carlo methods to forecast error statistics. *J. Geophys. Res.* **1994**, *99*, 10143–10162.
10. Julier, S.; Uhlmann, J.; Durrant-Whyte, H. A new approach for filtering nonlinear systems. In Proceedings of the American Control Conference, Seattle, WA, USA, 21–23 June 1995.
11. Julier, S.; Uhlmann, J.; Durrant-Whyte, H. A new method for the nonlinear transformation of means and covariances in filters and estimators. *IEEE Trans. Autom. Control* **2000**, *45*, 477–482.
12. Arulampalam, S.; Maskell, S.; Gordon, N.; Clapp, T. A tutorial on particle filters for on-line non-linear/non-Gaussian Bayesian tracking. *IEEE Trans. Autom. Control* **2002**, *30*, 174–189.
13. Chen, T.; Morris, J.; Martin, E. Particle filters for state and parameter estimation in batch processes. *J. Process Control* **2005**, *15*, 665–673.
14. Shenoy, A.V.; Prakash, J.; Prasad, V.; Shah, S.L.; McAuley, K.B. Practical issues in state estimation using particle filters: Case studies with polymer reactors. *J. Process Control* **2013**, *23*, 120–131.
15. Shao, X.; Huang, B.; Lee, J.M. Constrained Bayesian state estimation- A comparative study and a new particle filter based approach. *J. Process Control* **2010**, *20*, 143–157.
16. Crowley, T.J.; Meadows, E.S.; Kostoulas, E.; Doyle, F.D. Control of particle size distribution described by a population balance model of semibatch emulsion polymerization. *J. Process Control* **2000**, *10*, 419–432.
17. Kiparissides, C. Challenges in particulate polymerization reactor modeling and optimization: A population balance perspective. *J. Process Control* **2006**, *16*, 205–224.
18. Sajjadi, S.; Brooks, B.W. Unseeded semibatch emulsion polymerization of butyl acrylate: Bimodal particle size distribution. *J. Polym. Sci. A Polym. Chem.* **2000**, *38*, 528–545.
19. Doyle, F.J.; Soroush, M.; Cordeiro, C. Control of product quality in polymerization processes. In Proceedings of the Sixth International Conference on Chemical Process Control, Tucson, AZ, USA, 7–12 January 2001; AIChE Press: New York, NY, USA, 2002; pp. 290–306.
20. Flores-Cerrillo, J.; MacGregor, J.F. Control of particle size distributions in emulsion semibatch polymerization using mid-course correction policies. *Ind. Eng. Chem. Res.* **2002**, *41*, 1805–1814.
21. Bengtsson, T.; Snyder, C.; Nychka, D. Toward a nonlinear ensemble filter for high-dimensional systems. *J. Geophys. Res.* **2003**, *108*, 35–45.
22. Smith, K.W. Cluster ensemble Kalman filter. *Tellus* **2007**, *59A*, 749–757.
23. Dovera, L.; Della Rossa, E. Multimodal ensemble Kalman filtering using Gaussian mixture models. *Comput. Geosci.* **2011**, *15*, 307–323.
24. Dempster, A.P.; Laird, N.M.; Rubin, D.B. Maximum likelihood from incomplete data via the EM algorithm. *J. R. Stat. Soc. B* **1977**, *39*, 1–38.
25. Ormoneit, D.; Tresp, V. Improved Gaussian mixture density estimates using Bayesian penalty terms and network averaging. In *Advances in Neural Information Processing Systems 8*; Touretzky, D.S., Tesauro, G., Leen, T.K., Eds.; MIT Press: Cambridge, MA, USA, 1996; pp. 542–548.

26. Ueda, N.; Nakano, R.; Ghahramani, Z.; Hinton, G.E. SMEM algorithm for mixture models. *Neural Comput.* **2000**, *12*, 2109–2128.

27. Schwarz, G. Estimating the dimension of a model. *Ann. Stat.* **1978**, *6*, 461–464.

28. Hu, X.; Xu, L. Investigation on several model selection criteria for determining the number of cluster. *Neural Inf. Process.-Lett. Rev.* **2004**, *4*, 1–10.

29. Silva-Beard, A.; Flores-Tlacuahuac Silva, A. Effect of process design/operation on the steady-state operability of a methyl methacrylate polymerization reactor. *Ind. Eng. Chem. Res.* **1999**, *38*, 4790–4804.

30. Shenoy, A.V.; Prasad, V.; Shah, S.L. Comparison of unconstrained nonlinear state estimation techniques on a MMA polymer reactor. In Proceedings of the 9th International Symposium on Dynamics and Control of Process Systems (DYCOPS 2010), Leuven, Belgium, 5–9 July 2010.

31. Bavdekar, V.A.; Shah, S.L. Computing point estimates from a non-Gaussian posterior distribution using a probabilistic k-means clustering approach. *J. Process Control* **2014**, *24*, 487–497.

Modeling and Optimization of High-Performance Polymer Membrane Reactor Systems for Water–Gas Shift Reaction Applications

Andrew J. Radcliffe, Rajinder P. Singh, Kathryn A. Berchtold and Fernando V. Lima

Abstract: In production of electricity from coal, integrated gasification combined cycle plants typically operate with conventional packed bed reactors for the water-gas shift reaction, and a Selexol process for carbon dioxide removal. Implementation of membrane reactors in place of these two process units provides advantages such as increased carbon monoxide conversion, facilitated CO_2 removal/sequestration and process intensification. Proposed H_2-selective membranes for these reactors are typically of palladium alloy or ceramic due to their outstanding gas separation properties; however, on an industrial scale, the cost of such materials may become exorbitant. High-performance polymeric membranes, such as polybenzimidazoles (PBIs), present themselves as low-cost alternatives with gas separation properties suitable for use in such membrane reactors, given their significant thermal and chemical stability. In this work, the performance of a class of high-performance polymeric membranes is assessed for use in integrated gasification combined cycle (IGCC) units operated with carbon capture, subject to constraints on equipment and process streams. Several systems are considered for use with the polymeric membranes, including membrane reactors and permeative stage reactors. Based upon models developed for each configuration, constrained optimization problems are formulated which seek to more efficiently employ membrane surface area. From the optimization results, the limiting membrane parameter for achieving all carbon capture and H_2 production specifications for water–gas shift reactor applications is determined to be the selectivity, α_{H_2/CO_2}, and thus a minimum value of this parameter which satisfies all the constraints is identified for each analyzed configuration. For a CO_2 capture value of 90%, this value is found to be $\alpha = 61$ for the membrane reactor and the 3-stage permeative stage reactor and $\alpha = 62$ for the 2-stage permeative stage reactor. The proposed systems approach has the potential to be employed to identify performance limitations associated with membrane materials to guide the development of future polymeric and other advanced materials with desired membrane characteristics for energy and environmental applications.

Reprinted from *Processes*. Cite as: Radcliffe, A.J.; Singh, R.P.; Berchtold, K.A.; Lima, F.V. Modeling and Optimization of High-Performance Polymer Membrane Reactor Systems for Water–Gas Shift Reaction Applications. *Processes* **2016**, *4*, 8.

1. Introduction and Prior Work

As the world transitions to a more environmentally conscious economy, the importance of hydrogen (H_2) production processes is paramount. Hydrocarbons such as petroleum, natural gas, coal and biomass serve as the principal sources of H_2, which will see use as a feedstock in myriad clean energy and chemical production processes. As H_2 production from hydrocarbons generates carbon dioxide (CO_2), processes incorporating carbon capture technologies are necessary to achieve the objective of reduction of CO_2 emissions in accordance with protocols that seek to mitigate global climate change. Based upon extrapolation of the rates of consumption and available reserves, projections posit that petroleum resources may be depleted within 50 years and natural gas resources within 100; however, coal resources may exhibit their current availability for a couple hundred years [1]. Consequently, emerging energy technologies that utilize coal as the feedstock, such as integrated gasification combined cycle (IGCC) power plants operated with carbon capture, are particularly promising.

Coal-based IGCC units produce electricity through a synthesis gas (syngas) intermediate, which is subjected to the water-gas shift (WGS) reaction to maximize H_2 produced prior to the stream being sent to the gas turbine portion of the unit. An IGCC process scheme with carbon capture typically utilizes packed-bed WGS reactors followed by CO_2 removal by a Selexol process [2,3]. An alternative to this method of syngas conversion utilizes membrane reactors (MRs) equipped with H_2-selective membranes, which grant advantages such as increased carbon monoxide (CO) conversion, facilitated CO_2 removal/sequestration (CO_2-rich effluent is produced at high pressure), and process intensification through a reduction to the total number of process units [2,4].

There are challenges inherent to the use of MRs for such an application as the H_2-selective membranes must be stable under high-temperature and extreme pressure conditions in the presence of water and contaminants such as hydrogen sulfide (H_2S). H_2-selective membranes commonly considered for this application are as follows: (i) zeolite-based molecular sieves; (ii) dense metals such as Pd; and (iii) polymeric membranes. Of these potential membrane materials, (i) and (ii) possess highly favorable gas separation properties in terms of selectivity and flux, but the cost for these materials may be prohibitive for industrial-scale application. Only some polymeric membranes can be considered for the WGS application, as the elevated operating temperature of the MR unit is often outside the stability limits of the membrane material or the membrane material exhibits limited gas separation properties at the operating temperatures defined by the WGS-MR. However, if the aforementioned performance and stability challenges are addressed, polymeric membranes possessing suitable gas separation properties offer the potential to greatly reduce the cost of industrial-scale-MR implementation.

Polybenzimidazoles (PBIs) represent one such class of high performance polymers having exceptional chemical and physical characteristics enabling H_2/CO_2 separation in challenging thermo-chemical environments. These materials exhibit molecular-sieving mechanisms analogous to those observed in zeolite-based membranes, which imbues these materials with attractive H_2/CO_2 selectivity for syngas separations. High-performance polymeric materials have also been found to exhibit good thermal stability up to 400 °C and chemical stability in the presence of common syngas contaminants [5,6].

One objective of this study is to assess the feasibility of the state-of-the-art high-performance polymeric materials for use in membrane reactor systems with respect to performance constraints set forth by the U.S. Department of Energy (DOE) for pre-combustion CO conversion/CO_2 separation processes within IGCC units [3]. In this study, the performance characteristics of PBI-based membranes, as demonstrated by Berchtold and coworkers in [5], are used to develop the benchmark case for the polymer membrane-based MR process schemes investigated and developed herein. These PBI-based membrane materials have demonstrated industrially attractive H_2/CO_2 separation characteristics including ideal H_2 permeabilities between 58 and 78 barrer and H_2/CO_2 selectivities between 23 and 43 at 250 °C [5,7]. Additionally, this study seeks to determine the minimum membrane characteristics needed to satisfy the DOE's performance constraints by considering process models for several reactor designs. The performance of the various reactors is assessed in the base case conformations, which are then modified by considering different catalyst/membrane placement about the axial axis. Alternative reactor designs are developed by seeking to maximize reactor performance (H_2 recovery) for the minimum reactor cost as determined by the required membrane surface area. As demonstrated previously, an optimization problem is formulated to guide these designs [2].

With regard to the systems analysis, there are several MR models (utilizing H_2-selective membranes) related to the WGS reaction available in the literature, encompassing the range from 1-D/isothermal to 2-D/non-isothermal. Also available in the literature are MR models that employ H_2-selective membranes relating to widely varied applications (see [2] for a summary for MR models, efforts and applications). A review of literature shows a few computational modeling studies based on membrane reactors employing polymeric membranes [8,9], due in part to the temperature limitations imposed by available polymers. However, the literature suggests a lack of studies on optimization of polymer-based MR configurations. Recent and continued development and demonstration of high performance polymers such as PBIs for potential use in challenging membrane separation environments, such as those encountered in the vicinity of the WGS reaction, presents an opportunity to derive a MR model for a system utilizing such H_2-selective polymers and subsequently evaluate their potential in this challenging separations role [5,7,10]. This study is focused on H_2-selective membranes due to their advantages over CO_2-selective membranes in IGCC process schemes, as discussed by [11].

Moreover, several optimization studies relating to packed-bed MRs and reactor systems employing membrane separators are available in the literature. These studies have utilized H_2-selective membranes (ceramic or Pd) to formulate optimization problems that examine staged membrane reactors [12–14] and traditional MRs [2]. In the case of the staged membrane reactors, the optimization problems were formulated with the objective of maximizing methane conversion, H_2 recovery or H_2 yield in a steam methane reforming (SMR) process employing a Pd-based membrane. These studies considered a permeative stage membrane reactor (PSMR) with a fixed number of stages, or a staged membrane reactor (continuous membrane, catalyst packing with inert stages). The decision variables were composed of the catalyst/membrane stage lengths, but the problem was not subject to performance constraints. For the case of the traditional MR performing the WGS reaction, the study formulated an optimization problem in terms of economic variables that maximizes performance (H_2 recovery) for the minimum cost (membrane surface area) subject to multiple constraints on reactor effluent streams by considering alternative catalyst/membrane placement about the axial axis of the reactor. With regard to the available literature, it is worth noting that computational studies of SMR or WGS processes that use Pd/micro-porous ceramic membranes have H_2 selectivity values that are comparatively larger than those of polymeric membranes.

Thus, the computational study performed here of MR systems employing novel polymeric membrane materials provides insight into their feasibility for WGS reaction applications. Additionally, such a study may be used to identify performance limitations associated with the material, which may be used to guide the development of future polymeric materials with desired membrane characteristics. To this end, mathematical models are developed for traditional MRs and PSMRs using the performance characteristics of PBI membranes; these models are subsequently employed to develop reactor designs that satisfy the set of performance constraints set forth by the U.S. DOE for pre-combustion CO conversion (WGS reaction)/CO_2 separation processes within IGCC units. Using these process models for the MR and PSMR cases, constrained optimization problems are formulated that seek to maximize performance (H_2 recovery) through minimization of membrane surface area—this is achieved by considering alternate membrane placement about the

158

axial axis of the reactor. Through the formulation of two optimization problems, the performance-limiting membrane parameter is identified and a minimum value that satisfies all equipment/stream constraints is successfully calculated for each configuration. This study contributes insight into identifying and prioritizing the membrane parameters that should be the focus of future polymeric membrane development efforts, and provides a minimum value for key parameters that satisfy the set of six performance constraints; to this end, it is worth noting that the minimum selectivity value (one such key parameter) presented here is unique to the operating temperature and pressure of the process units, and the syngas feed/steam sweep flowrates.

2. Systems Analysis—Process Modeling, Simulation and Optimization Approach

2.1. Membrane Modeling

The membrane reactor model employed for the performance assessment and optimization studies is a one-dimensional, isothermal model in which operation is steady-state and the ideal gas law is assumed to hold. The 1-D and isothermal model assumptions are reasonable for a laboratory-scale membrane reactor [15]. This model was developed based on the WGS-MR model in [2]. A summary of the development is presented below; refer to [2] for additional detail. Assuming plug-flow operation, the membrane reactor model consists of species mole balances for co-current and counter-current cases:

Mole balance, tube:

$$\frac{dF_{i,t}}{dz} = r_i A_t - J_i \pi d_t$$

where $F_{i,t}$ is the flow rate in the tube, r_i is the species reaction rate, A_t is the cross-sectional area of the tube, J_i is the molar flux across the tube wall, and d_t is the tube diameter. Additionally, $r_i = r_{CO}$ for i=CO, H_2O; $r_i = -r_{CO}$ for i = CO_2, H_2 and $r_i = 0$ for i = N_2. The reaction rate, r_{CO}, is the rate associated with the Cu/ZnO/Al$_2$O$_3$ catalyst [16].

Mole balance, shell:

$$(\pm) \frac{dF_{i,s}}{dz} = J_i \pi d_t$$

where $F_{i,s}$ is the flow rate in the shell. The positive coefficient corresponds to co-current operation and negative to counter-current. For the permeative stage membrane reactor, the only differences in the model are: $J_i = 0$ in the reactor stages, and $r_i = 0$ in the membrane separator stages. The resulting mathematical model consists of an ODE system corresponding to an initial value problem (co-current) or a boundary value problem (counter-current), both of which may be solved using the MATLAB subroutines *ode15s* or *bvp4c*, respectively. A schematic of the

159

counter-current MR design employed here is shown in Figure 1; co-current operation of the unit would align the feed/sweep in the same direction with respect to the axial axis. A comparison using models developed in Aspen Plus considering the scenarios of an IGCC plant with the WGS-MR process against the CO shift followed by physical absorption (e.g., Selexol) technologies for CO_2 capture will be analyzed in a future publication.

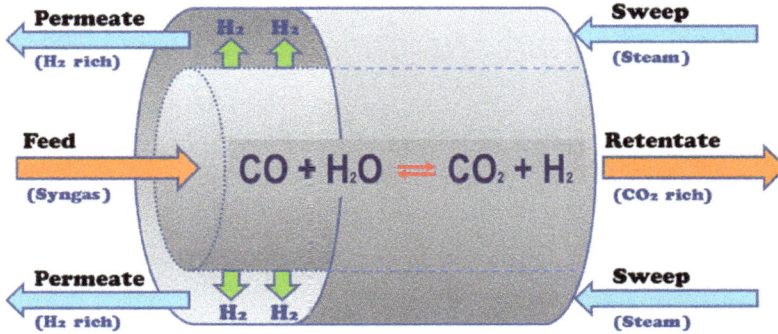

Figure 1. The membrane reactor consists of a shell and tube setup in which the tube is packed with catalyst and the membrane is fixed to the tube wall; reaction/permeation occur simultaneously.

The flux through the high-performance polymer membrane is assumed to be Fickian activated diffusion, which is proportionate to the component partial pressure difference across the membrane [17]; and is described by:

$$J_i = Q_i \Delta p_i$$

where Δp_i is the partial pressure difference of the component across the membrane. The permeance of a component i, Q_i, is determined by the membrane properties that are taken from test systems available in the literature. The permeability (P_i) of a component through polymers is considered to be the product of the diffusion coefficient, D_i, and solubility coefficient, S_i [18].

Noting that selectivity (α) of species i to j is defined as the ratio of the permeability of species i to j, such that:

$$\alpha_{ij} = \frac{P_i}{P_j}$$

and the permeance of a species is:

$$Q_i = \frac{P_i}{\delta_{mem}}$$

where δ_{mem} is the membrane thickness.

In this paper, we focus on a class of high-performance PBI-based H_2-selective polymeric membranes utilizing their demonstrated separation performance characteristics in multiple platforms including flat sheet, tubular, and hollow fiber [5,7,10] for an assumed industrially relevant selective layer thickness range of 100-200 nm. Thus, the following membrane characteristics are used in this study:

- Q_{H_2} = 250 GPU
- α_{H_2/CO_2} = 20–28

Also, for this study we assume that at high temperatures, these membranes have high permeability to water (α_{H_2/H_2O} = 0.33) and low permeability to the other species considered here. The H_2/CO selectivity in this study was assumed to be 99, similar to experimentally measured H_2/N_2 selectivity of 99. However, based on the size difference between CO (kinetic diameter = 3.76 Å) and N_2 (3.64 Å), a H_2/CO selectivity greater than 99 would be possible. In particular, both H_2/CO_2 and H_2/CO permselectivities must be high for this application as one desires to produce a purified CO_2 effluent from the reactor side.

2.2. Simulation Set Up

The reactor feed composition/molar flow rate and sweep composition/molar flow rate are drawn from [2]; the feed corresponds to a syngas stream from the gasifier after steam injection (it is assumed that sulfurous compounds and other impurities have been removed) while the sweep composition is pure steam. The feed/sweep compositions are summarized in Table 1.

Table 1. Molar composition of reactor inlet streams, given in mole fraction.

Component	Syngas Feed	Steam Sweep
H_2	0.1933	0
CO_2	0.0568	0
H_2O	0.4886	1
CO	0.2443	0
N_2	0.017	0

The performance of each reactor system is evaluated in terms of the three performance goals set forth by the U.S. DOE for CO conversion, H_2 recovery and CO_2 capture in addition to three constraints on the reactor effluent streams as defined [2]:

- CO conversion (X_{CO})

$$X_{CO} = \frac{CO\,converted}{CO\,in\,feed} = \frac{F_{CO,f} - (F_{CO,r} + F_{CO,p})}{F_{CO,f}} \geqslant 98\%$$

- H_2 recovery (R_{H_2})

$$R_{H_2} = \frac{H_2\,in\,permeate}{(H_2 + CO)\,in\,feed} = \frac{F_{H_2,p}}{F_{H_2,f} + F_{CO,f}} \geqslant 95\%$$

- CO_2 Capture (C_{CO_2})

$$C_{CO_2} = \frac{carbon\,in\,retentate}{carbon\,in\,feed} = \frac{F_{CO,r} + F_{CO_2,r}}{F_{CO,f} + F_{CO_2,f}} \geqslant 90\%$$

- $CO_2 + H_2O$ purity in the retentate

$$purity_{CO_2+H_2O,r} \geqslant 95\%$$

- H_2 mole fraction in the retentate

$$y_{H_2,r} \leqslant 4\%$$

- H_2 purity in the permeate

$$purity_{H_2,p} \geqslant 44\%$$

The reactor designs considered in the performance assessment and the optimization problems (as the initial guess) are a 2-stage PSMR, 3-stage PSMR and a conventional MR. Reactor feed/sweep molar flow rate, composition and flow arrangement (counter-current) are kept constant across all simulations, as are all other reactor operating conditions such as temperature (constant at 300 °C), tube/shell pressure (47.63 atm/25.86 atm, respectively), mass of catalyst (20 mg), and tube/shell diameter (1.02 cm/6.12 cm, respectively). Flow arrangement was fixed as counter-current as co-current results were consistently unable to satisfy any constraint other than CO conversion; this result was observed in [2] and was verified in the co-current simulations performed as part of this study. With regard to the membrane properties, maintenance of constant temperature, and fixed permeance (and selectivity) values for the analyzed polymer material are considered. For base case performance studies, total reactor length is kept constant at 300 cm (thereby making membrane surface area a constant) for MRs, and for the PSMRs the catalyst/membrane are divided into their components and arranged in equally sized pieces such that the total length (L) of the 2/3-stage PSMRs are 600 cm—this corresponds to catalyst/membrane lengths of 150 cm per stage in the 2-stage PSMR and 100 cm per stage in the 3-stage PSMR.

2.3. Optimization Problem Formulations

The first of the two formulated optimization problems seeks to maximize reactor performance, expressed by H_2 production, subject to the six performance constraints, by considering alternative catalyst/membrane placement while minimizing the total membrane surface area (S_m) required. To this end, cost parameters were associated with R_{H_2} and S_m in accordance with the method set forth by [2]. The main difference in this analysis is the cost of the high-performance polymer membrane. The polymer membrane cost is estimated to range between \$5–200/m^2 depending on the membrane module platform [19], where an all polymeric hollow fiber platform typically provides the best economics, *i.e.*, lowest cost, and a porous inorganic supported composite tubular membrane platform is typically the highest cost option. The application of robust stainless steel porous supports with weldability and correspondingly perceived lower risk for incorporation into MR configuration can further increase the cost of the resulting polymer/inorganic tubular membranes [20]. For the purposes of this work, we have chosen the highly robust tubular membrane platform as our benchmark. As such, a cost of $\$_m$ = \$500/m^2 has been assumed as an upper bound estimate for the cost of the polymer membrane selective layer on a tubular stainless steel support (the platform utilized by [5] in their year plus evaluations of this membrane in elevated temperature separation environments).

The objective function formulated for this optimization problem is defined as:

$$\varphi = \min_x \left[cost_m - credit_{H_2} \right]$$

In which:

$$credit_{H_2} = F_{H_2,p} HHV_{H_2} \$_{H_2} Op$$

$$cost_{m,2-\text{stage PSMR}} = \$_m \pi d_t \left(l_2 - l_1 + l_4 - l_3 \right)$$

$$cost_{m,3-\text{stage PSMR}} = \$_m \pi d_t \left(l_2 - l_1 + l_4 - l_3 + l_6 - l_5 \right)$$

$$cost_{m,\text{MR}} = \$_m \pi d_t \left(l_6 - l_5 + l_8 - l_7 + L - l_9 \right)$$

where $F_{H_2,p}$(mol/s) is the molar flow rate of H_2 in the permeate, HHV_{H_2} (BTU/mol) is the higher heating value of H_2, $\$_{H_2}$ (\$/BTU) is the monetary credit associated with the heating value, and Op is the 1-year operating period in seconds.

In particular, for the 2-stage PSMR design depicted in Figure 2, the vector of decision variables is as follows:

$$x_{2-\text{stage PSMR}} = \left[l_1 \ l_2 \ l_3 \ l_4 \right]^T$$

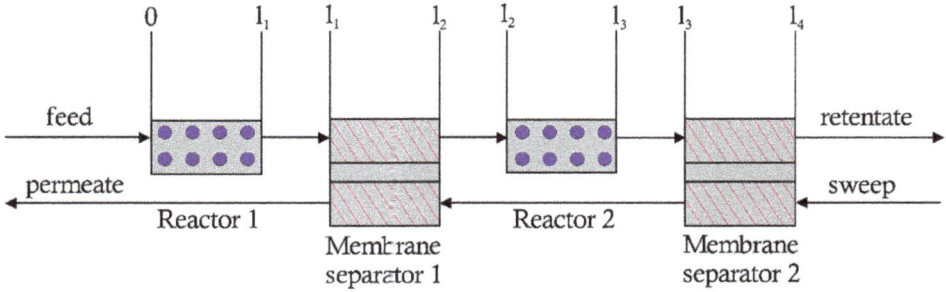

Figure 2. Arrangement of decision variables about the axial axis of the 2-stage permeative stage membrane reactor (PSMR).

Subject to the dimensional constraints on the catalyst stage lengths:

$$l_1 > 0, \; l_3 \geqslant l_2$$

and for the membrane stage lengths:

$$l_2 \geqslant l_1, \; l_4 \geqslant l_3, \; l_4 \leqslant L$$

The initial guess for the 2-stage PSMR optimization problem corresponds to four equally sized stages (two catalyst, two membrane) consisting of the same membrane surface area and catalyst mass as the conventional MR, given by:

$$x_{2-\text{stage PSMR,initial}} = [150 \; 300 \; 450 \; 600]^T$$

The vector of decision variables for the 3-stage PSMR design shown in Figure 3 is as follows:

$$x_{3-\text{stage PSMR}} = [l_1 \; l_2 \; l_3 \; l_4 \; l_5 \; l_6]^T$$

Subject to the dimensional constraints on the catalyst stage lengths:

$$l_1 > 0, \; l_3 \geqslant l_2, \; l_5 \geqslant l_4$$

and for the membrane stage lengths:

$$l_2 \geqslant l_1, \; l_4 \geqslant l_3, \; l_6 \geqslant l_5, \; l_6 \leqslant L$$

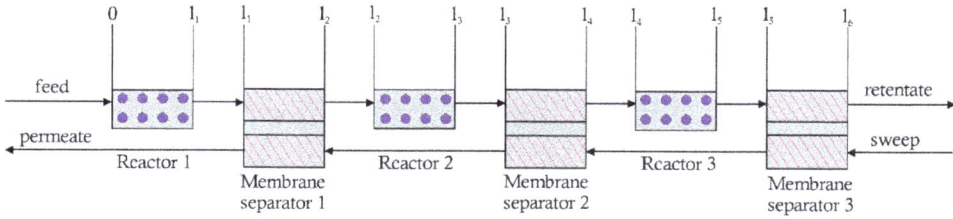

Figure 3. Arrangement of decision variables about the axial axis of a 3-stage PSMR.

The initial guess for the 3-stage PSMR corresponds to six equally sized stages of catalyst/membrane, using the same membrane surface area and catalyst mass as the 2-stage PSMR, is as follows:

$$x_{3-\text{stage PSMR,initial}} = [100\ 200\ 300\ 400\ 500\ 600]^T$$

The vector of decision variables corresponding to the MR design presented in Figure 4 is as follows:

$$x_{\text{MR}} = [l_1\ l_2\ l_3\ l_4\ l_5\ l_6\ l_7\ l_8\ l_9]^T$$

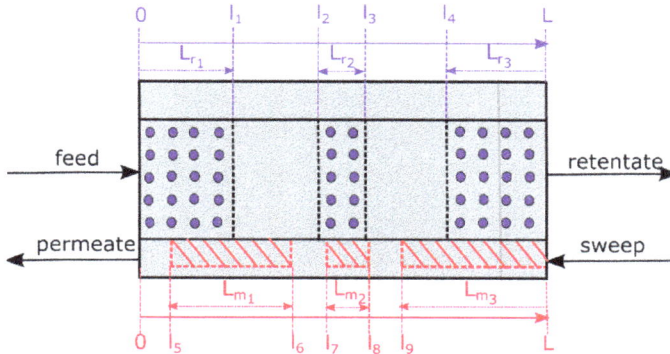

Figure 4. Arrangement of decision variables about the axial axis of the MR.

Subject to the dimensional constraints on the reaction zone:

$$l_1 > 0,\ l_2 \geqslant l_1,\ l_3 \geqslant l_2,\ l_4 \geqslant l_3,\ l_4 \leqslant L$$

and in the permeation zone:

$$l_5 \geqslant 0,\ l_6 \geqslant l_5,\ l_7 \geqslant l_6,\ l_8 \geqslant l_7,\ l_9 \geqslant l_8,\ l_9 \leqslant L$$

The initial guess for the MR is a conventional case in which catalyst/membrane are present along the whole axial length, the vector for which is:

$$x_{MR,initial} = [100\ 100\ 200\ 200\ 0\ 100\ 100\ 200\ 200]^T$$

A second optimization problem is formulated to verify the hypothesis that minimization of membrane surface area also corresponds to maximization of the limiting performance parameter (C_{CO_2}), using the five remaining nonlinear constraints on reactor performance with the same linear constraints as the cost optimization problem described above. The difference in this case is in the formulation of the objective function, which is defined as:

$$\varphi = \min_{x} \left[-C_{CO_2} \right]$$

As both optimization problems possess nonlinear objective functions subject to a set of nonlinear constraints, solutions may be obtained through the MATLAB *fmincon* subroutine employing the *"active-set"* algorithm.

3. Systems Analysis Results

3.1. 2-Stage Permeative Stage Membrane Reactor Performance, Optimization

The performance of the high-performance polymeric membrane is first assessed as part of a 2-stage PSMR, using permeance (Q_{H_2} = 250 GPU) and H_2/CO_2 selectivity (α_{H_2/CO_2} = 28) values. The results of this simulation are summarized in Table 2, which employs the feed/sweep flow rate, flow composition, flow arrangement (counter-current), temperature and pressure conditions as defined above. The placement of catalyst/membrane correspond to $x_{2-stage\ PSMR,initial}$ or the base case design.

Table 2. Performance of polymer membrane in a 2-stage PSMR (Q_{H_2} = 250 GPU, α_{H_2/CO_2} = 28).

Parameter	Value (%)	Target (%)
X_{CO}	98.95	98
R_{H_2}	98.60	95
C_{CO_2}	74.71	90
$purity_{CO_2+H_2O,r}$	95.98	95
$purity_{H_2,F}$	41.71	44
$y_{H_2,r}$	0.77	$\leqslant 4$

Thus, two stages for this base case satisfies all but carbon capture and H_2 purity constraints for this reactor configuration. Also through simulations, the

166

limiting membrane characteristic for these performance parameters is identified as the H_2/CO_2 selectivity (though reduction to total membrane surface area improves carbon capture, sufficient reductions cannot be performed should one desire to satisfy the remaining five performance constraints). Thus, an incremental variation of H_2/CO_2 selectivity is performed for the range of $\alpha = 25$–75 with the objective of determining the minimum H_2/CO_2 selectivity that would satisfy the carbon capture and permeate hydrogen purity constraints in an optimized (minimum membrane surface area) 2-stage PSMR. Utilizing an α increment of 5, eleven optimization problems were formulated and solved using the technique described above. Designs employing values of α_{H_2/CO_2} greater than 30 satisfy all but the carbon capture constraint; the first design to satisfy all six constraints in a 2-stage PSMR falls within the range $\alpha_{H_2/CO_2} = 60$–65, and occurs at a H_2/CO_2 selectivity value of approximately 62. As the solutions represent a maximization of carbon capture attainable in a 2-stage PSMR, while satisfying the other five performance constraints, the results may be used to determine the minimum H_2/CO_2 selectivity needed to satisfy a given carbon capture constraint. The carbon capture resulting from varying selectivity on the range of 25–75 in an optimized 2-stage PSMR employing the minimum membrane area is shown in Figure 5.

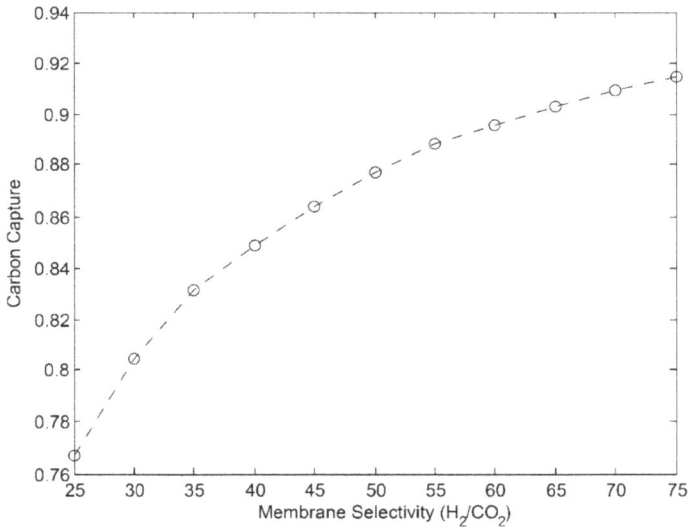

Figure 5. Maximization of carbon capture in an optimized 2-stage PSMR; only the design corresponding to $\alpha_{H_2/CO_2} = 25$ fails to satisfy $purity_{H_2,p}$ otherwise the remaining five constraints are satisfied for all cases; $C_{CO_2} \geqslant 90\%$ corresponds to the target value used in this study.

The solution vector for the case corresponding to $\alpha_{H_2/CO_2} = 60$ is shown below; this case is closest to satisfying $C_{CO_2} \geq 0.90$ in the 2-stage PSMR.

$$x_{2-\text{stage PSMR,final,}\alpha \frac{H_2}{CO_2}=60} = [156.28 \ 295.85 \ 454.85 \ 579.93]^T$$

Upon examination of the optimization results representing a minimization of membrane surface area, a pattern is noted in the optimal 2-stage PSMR catalyst/membrane placement. In each design, there is an approximately equal distribution of catalyst mass across the two stages; the membrane stages are unevenly distributed, with the first membrane stage slightly larger than the second for all cases. The membrane placement likely results from the relatively high partial pressure of H_2 in the stream exiting the first reactor stage (see Figure 6), thus, placement of more membrane directly after the first reactor allows for greater utilization of the permeation driving force and consequently greater H_2 recovery in this region (relative to the second membrane separator stage). Figure 7 shows the profiles for the H_2 reaction and diffusion rates as function of axial axis for this optimized 2-stage PSMR.

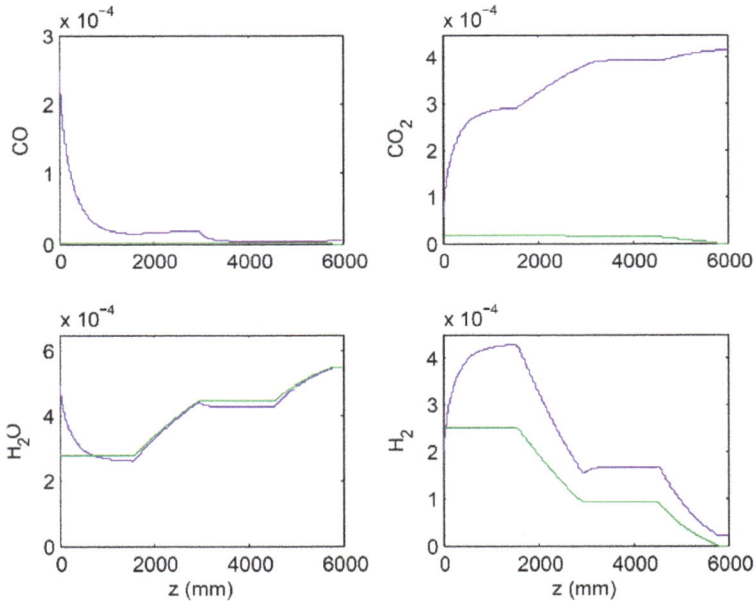

Figure 6. Species concentration (mol/cm^3) as function of axial length for optimized 2-stage PSMR for $\alpha_{H_2/CO_2} = 60$, corresponding to the solution vector shown above; solid blue and solid green lines denote feed and sweep streams, respectively.

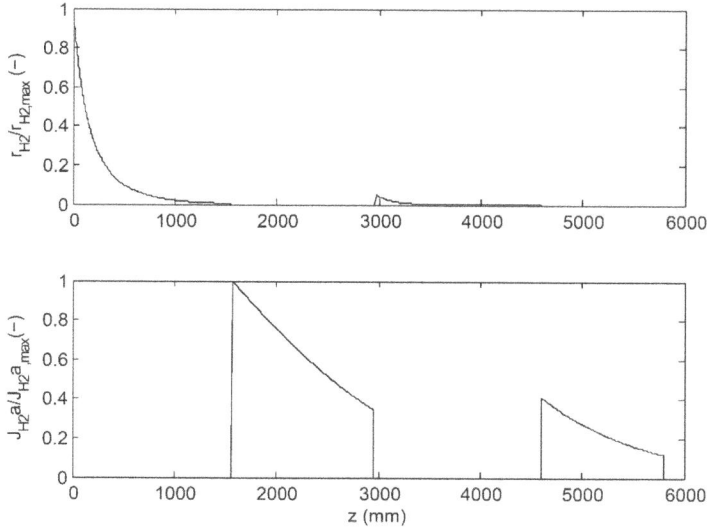

Figure 7. H_2 reaction (r_{H_2}) and diffusion ($J_{H_2}a$, $a \equiv 4/d_t$) rates as function of axial axis for the optimized 2-stage PSMR ($\alpha_{H_2/CO_2} = 60$); dimensionless quantities scaled by the maximum H_2 reaction and diffusion rates, respectively.

3.2. 3-Stage Permeative Stage Membrane Reactor Performance, Optimization

The performance of the polymeric membrane is assessed next as a 3-stage PSMR, using $Q_{H_2} = 250$ GPU, and $\alpha_{H_2/CO_2} = 28$ once again as base case. The results of this simulation, which uses precisely the same process conditions as the 2-stage PSMR, are summarized in Table 3; the placement of catalyst/membrane in the 3-stage PSMR correspond to $x_{3-stage\ PSMR,initial}$.

Table 3. Performance of polymeric membrane in a 3-stage PSMR ($Q_{H_2} = 250$ GPU, $\alpha_{H_2/CO_2} = 28$).

Parameter	Value (%)	Target (%)
X_{CO}	99.44	98
R_{H_2}	98.82	95
C_{CO_2}	74.73	90
$purity_{CO_2+H_2O,r}$	96.17	95
$purity_{H_2,p}$	41.82	44
$y_{H_2,r}$	0.71	$\leqslant 4$

Once more, given the membrane properties, the material did not to satisfy the carbon capture/permeate hydrogen purity constraint while satisfying the other four performance constraints. Similar to the previous case, an incremental variation of

169

H_2/CO_2 selectivity was performed for the range of $\alpha = 25$–75 so as to determine the minimum H_2/CO_2 selectivity that would satisfy the carbon capture and permeate hydrogen purity constraints in an optimized 3-stage PSMR. These simulations indicate that designs employing H_2/CO_2 selectivity values greater than 30 satisfy all but the carbon capture constraint, with the first design satisfying all six constraints falling in the α_{H_2/CO_2} range of 60–65 with the minimum H_2/CO_2 selectivity that satisfies all constraints of approximately 61. The carbon capture resulting from varying selectivity on the range of 25–75 in an optimized 3-stage PSMR employing the minimum membrane area is shown in Figure 8, along with the previously obtained result for the 2-stage reactor.

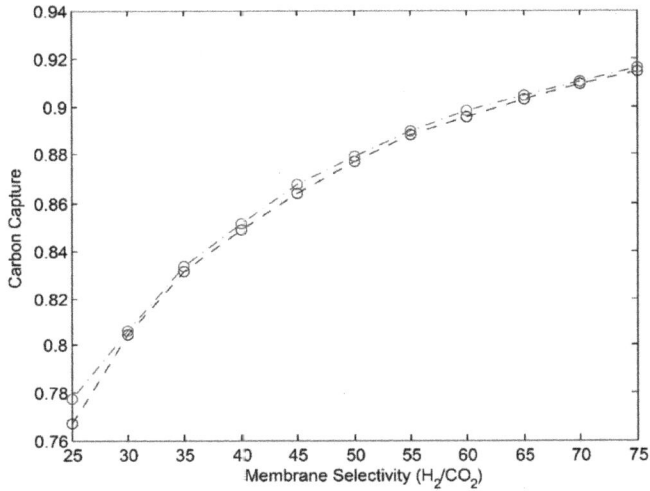

Figure 8. Maximization of carbon capture in an optimized 3-stage PSMR (blue) as well as optimized 2-stage PSMR (black); only the designs corresponding to $\alpha_{H_2/CO_2} = 25$ fail to satisfy $purity_{H_2,p}$, otherwise the remaining five constraints are satisfied for all cases; $C_{CO_2} \geqslant 90\%$ corresponds to the target value used in this study.

The solution vector for the case corresponding to $\alpha_{H_2/CO_2} = 60$ is shown below; this case is closest to satisfying $C_{CO_2} \geqslant 0.90$ in the 3-stage PSMR.

$$x_{3-\text{stage PSMR,final,}\alpha \frac{H_2}{CO_2}=60} = [107.26\ 192.26\ 309.39\ 399.86\ 499.83\ 583.20]^{T}$$

From the optimization results corresponding to a minimization of membrane area for the given range of selectivity, the designs conformed to a general pattern (as viewed from left to right in Figure 3): an uneven catalyst distribution that preferentially placed the most catalyst in the second reactor stage, and slightly more catalyst in the first reactor stage than the third reactor stage (in this case differences in catalyst amount corresponded to less than 10% of the total catalyst mass). As for the membrane area placement, a pattern was also observed across the range of selectivity values in which more membrane was utilized in the second stage than the first/third stages; however, the difference in total membrane area between each stage was small (the second membrane separator stage utilized 5%–15% more membrane than the first/third stages). The difference in membrane placement is likely due to the relatively large difference in partial pressure presented in the second membrane stage (see Figure 9); at this point, the sweep gas has a relatively low H_2 mole fraction while the reactor stream has a relatively larger H_2 mole fraction (having passed over approximately $2/3$ of the total catalyst mass (two reactor stages), but only one membrane separator stage). When comparing the performance of the 2-stage PSMR to that of the 3-stage PSMR, it is worth noting that the use of three reaction/permeation stages allowed for increased CO conversion and H_2 recovery by relieving equilibrium limitations on the WGS reaction (the first membrane separator is implemented after the reactor feed sees approximately $1/3$ of the total catalyst, rather than ½ the total catalyst as in the 2-stage PSMR). Additionally, the 3-stage design permitted increases to carbon capture through more efficient membrane utilization (optimized 3-stage PSMRs employed slightly less membrane area than optimized 2-stage PSMRs). Figure 10 depicts the profiles for the H_2 reaction and diffusion rates as function of axial axis for this optimized 3-stage PSMR.

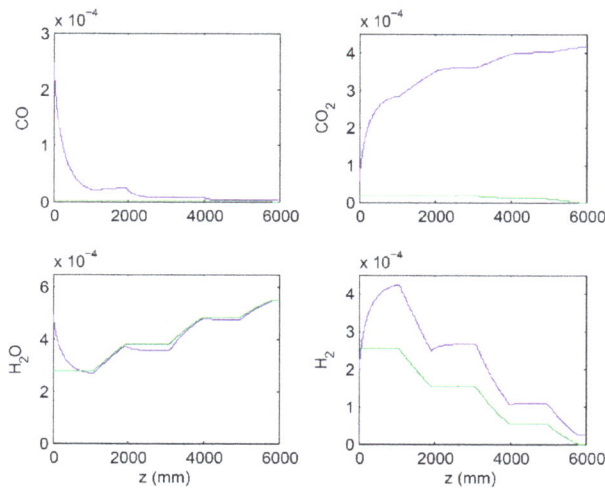

Figure 9. Species concentration (mol/cm^3) profiles in the optimized 3-stage PSMR for $\alpha_{H_2/CO_2} = 60$, corresponding to the solution vector shown above; solid blue and solid green lines denote feed and sweep streams, respectively.

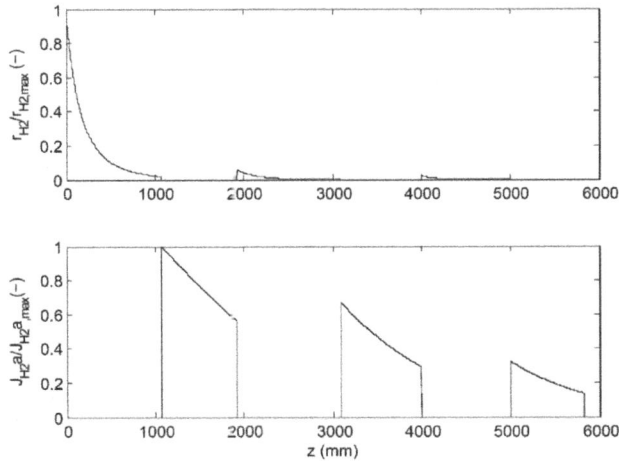

Figure 10. H_2 reaction (r_{H_2}) and diffusion ($J_{H_2}a$, $a \equiv 4/d_t$) rates for the optimized 3-stage PSMR ($\alpha_{H_2/CO_2} = 60$); dimensionless quantities scaled by the maximum H_2 reaction and diffusion rates, respectively.

3.3. Membrane Reactor Performance, Optimization

Finally, the performance of the polymeric membrane ($Q_{H_2} = 250$ GPU, $\alpha_{H_2/CO_2} = 28$) is assessed as a traditional membrane reactor. The results of this simulation, which uses the same process conditions as the 2,3-stage PSMRs, are summarized in Table 4; the reactor design is that of a conventional MR, which corresponds to $x_{MR,initial}$.

Noting that the material did not satisfy the carbon capture and hydrogen purity constraints, the same procedure outlined above for the 2-stage and 3-stage PSMRs was performed for the MR; that is: incremental variation of H_2/CO_2 selectivity was performed for the range of $\alpha = 25–75$ so as to determine the minimum H_2/CO_2 selectivity that would satisfy the carbon capture and permeate hydrogen purity constraints in an optimized MR. All values of α_{H_2/CO_2} greater than 30 satisfied the permeate hydrogen purity constraint, and the carbon capture constraint is satisfied in the range of $\alpha_{H_2/CO_2} = 60–65$, occurring at a value of approximately 61. The carbon capture resulting from varying selectivity on the range of 25–75 in an optimized MR utilizing the minimum membrane area is shown in Figure 11.

As for the arrangement of the membrane about the axial axis of the reactor, the optimal solutions (representing a minimization of membrane surface area) were all of the same general form as shown in Figure 12 The cause for variation in total membrane surface area was due to differences in selectivity, with lower values of α_{H_2/CO_2} allowing for dilution of the permeate with CO_2, which improved H_2 recovery by lowering H_2 partial pressure in the shell.

172

Table 4. Performance of polymeric membrane ($Q_{H_2} = 250$ GPU, $\alpha_{H_2/CO_2} = 28$) in a conventional MR for the same conditions of the PSMR.

Parameter	Value (%)	Target (%)
X_{CO}	99.36	98
R_{H_2}	98.38	95
C_{CO_2}	75.77	90
$purity_{CO_2+H_2O,r}$	96.05	95
$purity_{H_2,p}$	42.05	44
$y_{H_2,r}$	1.01	$\leqslant 4$

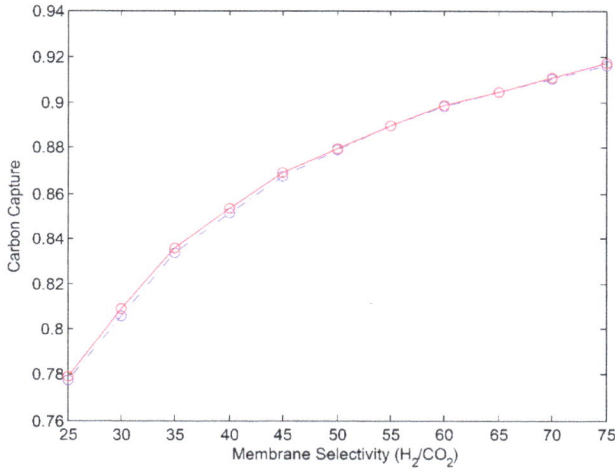

Figure 11. Maximization of carbon capture in an optimized MR (red) as well as 3-stage PSMR (blue); only the design corresponding to $\alpha_{H_2/CO_2} = 25$ fails to satisfy $purity_{H_2,p}$, otherwise the remaining five constraints are satisfied for all cases; $C_{CO_2} \geqslant 90\%$ corresponds to the target value used in this study.

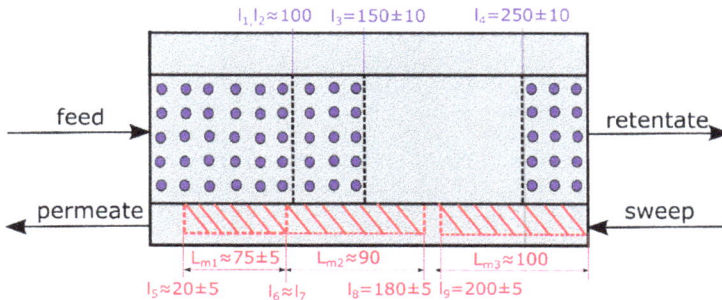

Figure 12. General optimized MR design; the solution specific to each H_2/CO_2 selectivity (on range 25–75) value falls within the range presented.

173

The solution vector for the case corresponding to $\alpha_{H_2/CO_2} = 60$ is shown below; this case is closest to satisfying $C_{CO_2} \geqslant 0.90$ in the MR (see Figure 13 for concentration profiles and Figure 14 for reaction/diffusion rate profiles associated with this case).

$$x_{MR,final,\alpha \frac{H_2}{CO_2}=60} = [98.43\ 98.43\ 142.53\ 257.30\ 15.22\ 90.54\ 96.88\ 185.21\ 200.20]^T$$

Each optimal design consisted of a short pre-shift zone lacking membrane; following the pre-shift zone, a region resembling a conventional MR exists until approximately 150 cm. From 150 cm to 250 cm, catalyst is absent and only membrane is placed so as to remove the reaction H_2 product from the tube side (though, there is a small section of membrane removed between 180 and 200 cm). From 250 cm to 300 cm, the design is once more that of the conventional MR, indicating that further CO conversion is best achieved after removal of a significant portion of the reaction products. The resulting design can be explained by the more efficient membrane area utilization resulting from increased H_2 partial pressure in the tube side achieved through use of a pre-shift followed by a conventional MR. In essence, the use of the combined pre-shift zone and conventional MR increases H_2 partial pressure in the reaction side, but as products build up the thermodynamic limitation associated with the WGS is increases. At this point, membrane is added to remove products and alleviate this limitation. Following the pre-shift zone in which product concentration is not sufficiently high, nearly continuous removal of product through the membrane permits increased CO conversion with respect to the staged reactors.

The optimized MRs exhibited slightly higher CO conversions than were obtained in the 2,3-stage PSMRs, as well as attained slightly higher H_2 recovery values; these results for $\alpha_{H_2/CO_2} = 60$ are summarized in Table 5.

Table 5. CO conversion and H_2 recovery values for optimized MR, 2-stage PSMR and 3-stage PSMR at $\alpha_{H_2/CO_2} = 60$.

Reactor Configuration	X_{CO} (%)	R_{H_2} (%)
2-stage PSMR	98.68	96.17
3-stage PSMR	99.16	96.07
MR	99.62	96.21

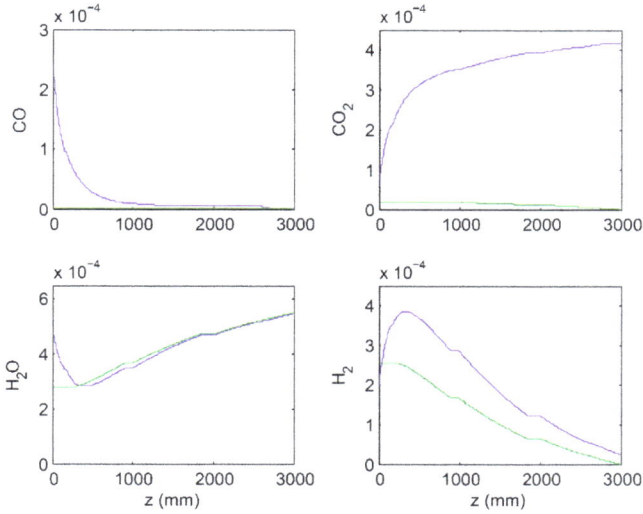

Figure 13. Species concentration (mol/cm^3) profiles as function of axial length in tube (solid blue line) and shell (solid green line) for the solution vector shown above (optimized MR with $\alpha_{H_2/CO_2} = 60$).

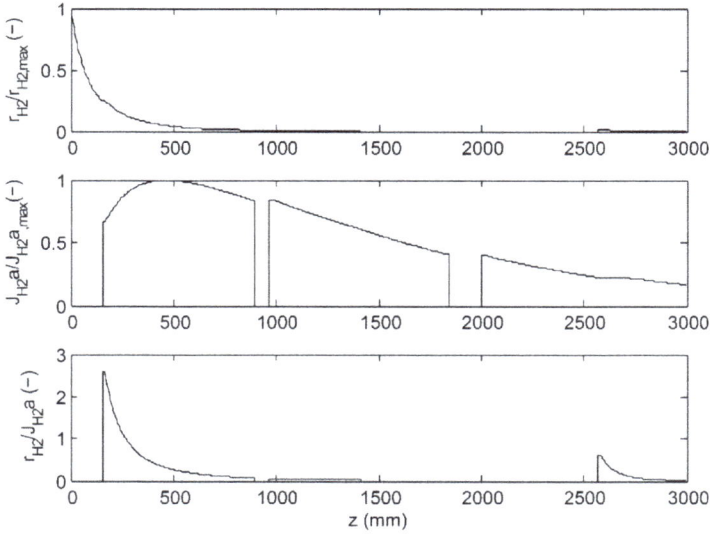

Figure 14. H_2 reaction (r_{H2}) and diffusion ($J_{H2}a$, $a \equiv 4/d_t$) rates, dimensionless quantities scaled by their maximum rates, for optimized MR ($\alpha_{H_2/CO_2} = 60$); the ratio of reaction to diffusion rate ($r_{H2}/J_{H2}a$, defined only where both catalyst and membrane are present) provides insight into the catalyst packing near the end of the reactor.

175

The increase in CO conversion is due to the continuous removal of products (CO_2, H_2) in portions of the MR, which constitutively relieves equilibrium limitations on the WGS reaction; the increase in H_2 recovery is due to the higher H_2 partial pressure differences between reactor/sweep streams achieved through selective placement of membrane in a reactor in which reaction/diffusion occur simultaneously. With regard to carbon capture, the MR achieves higher values at lower selectivity ($\alpha_{H_2/CO_2} \leqslant 50$) due to lower CO_2 partial pressure differences between the tube/shell (a positive factor for carbon capture, but this works against the operator for H_2 recovery); at higher values of selectivity, the 3-stage PSMR and MR produce nearly the same values for carbon capture, with the MR exceeding the 3-stage PSMR for α_{H_2/CO_2} = 55–75 by a very small margin.

From the 2-stage and 3-stage results, improved performance is the result of increasing the stage number, which allows for implementation of membrane stages at increased H_2 partial pressures (see Figure 6 *vs.* Figures 7 and 9 *vs.* Figure 10 for such configurations) As the stage number increases dramatically to the point at which the number of stages bring the reactor design to the limiting case (infinitesimally small stage lengths), the design equation associated with plug-flow becomes that associated with a continuous-flow stirred tank reactor (CSTR). As each stage operates as a CSTR and an infinite sequence of algebraic stages (infinite stage number reactor) may be taken to represent the differential reactor (the MR), large stage numbers cause the PSMR design to approach that of the MR, where the MR represents the maximum achievable performance for a given set of conditions. The trend for increasing stage number in PSMRs leading to operation resembling that of an MR is present as early as 2/3 stages for the case considered herein (Figures 8 and 11); four or more stages should yield results that increasingly resemble the MR. The ratio of reaction to diffusion rate in the membrane reactor configuration (see Figure 14) suggests that for cases in which minimization of membrane area is desired at fixed catalyst mass, more densely packing the catalyst (compared to spreading the fixed mass across the entire reactor length, as in the base case) can achieve this aim, provided temperature limitations are not present; alternatively for cases in which there is unlimited catalyst, dense packing throughout the entire reactor may serve to improve reactor performance.

In general, the supported tubular membrane platform benchmarked here is a desirable platform for the MR configuration as it allows efficient catalyst packing and the mechanical strength required to contain catalyst material. However, based on the results of this study, the performance of an optimized 3-staged PSMR can be comparable to that of a MR. This result presents an exciting opportunity for lower cost high performance hollow fiber membranes in this application. Given the high surface area to volume ratio of hollow fiber membrane modules and their resulting dramatically reduced containment vessel/module size, their cost per m^2 is estimated to be an order of magnitude lower than the tubular platform benchmarked here. As the optimization results presented here are sensitive to membrane cost, the influence of such membrane cost reduction opportunities on the process optimization will be explored in future studies.

4. Conclusions

One-dimensional isothermal models were developed for traditional MRs and PSMRs, and used to assess high-performance polymeric membrane reactor systems. Constrained cost optimization problems were formulated so as to systematically determine optimal reactor designs through more efficient membrane placement. As the solutions to these optimization problems also corresponded to a maximization of the limiting performance parameter, C_{CO_2}, an incremental search of H_2/CO_2 selectivity was then performed with the intent of determining C_{CO_2} as a function of α_{H_2/CO_2} at constant permeance. These designs were generated through the cost minimization optimization problem and the result that the economic optimum corresponds to maximization of carbon capture was verified by way of the second optimization problem formulation (utilizing identical decision variables), which sought to maximize C_{CO_2} subject to the other five performance constraints. This analysis was successfully completed for the three reactor designs considered (2-stage PSMR, 3-stage PSMR, MR). Graphs of carbon capture as a function of selectivity for fixed conditions were produced, which can guide the development of polymeric membrane materials to achieve all the desired specifications for their implementation in IGCC WGS environments.

Using the unit design framework considered herein, one may generalize from the process conditions of feed/sweep molar flow rate, flow composition/arrangement, reactor operating conditions (temperature, pressure), catalyst mass, tube dimensions and membrane properties to grant insight into future membrane material development by identifying the limiting parameter and determining a minimum value that satisfies all imposed constraints. Having identified a minimum value for a given parameter (α_{H_2/CO_2} in this case), a clear goal can be set for researchers in material development (should it be desired to use the process designs considered here). As the optimization results presented here are sensitive to membrane cost, it is desirable to investigate membranes with varied cost (*i.e.*, hollow fibers) to further understand the resulting outcomes in terms of required performance characteristics and optimized PSMR design. To that end, the presented modeling framework can be extended to evaluate performance of membrane materials in a systematic manner by considering several process designs in which material placement (catalyst, membrane) is guided by economic considerations and/or satisfaction of a set of performance constraints. The formulated optimization problem can also be extended to consider different operating conditions (temperature, pressure) for each reaction/separation module.

Acknowledgments: The authors (Andrew J. Radcliffe and Fernando V. Lima) would like to acknowledge West Virginia University for their financial assistance on this work. The authors (Rajinder P. Singh and Kathryn A. Berchtold) gratefully acknowledge the U.S. DOE/NETL-Strategic Center for Coal: Carbon Capture Program for financial support of the project under contract LANL-FE-308-13. Los Alamos National Laboratory is operated by Los Alamos National Security, LLC for DOE/NNSA under Contract DE-AC52-06NA25396. Additionally, the authors would like to thank Juan Carlos Carrasco for his assistance on illustrations.

Author Contributions: This paper is a collaborative work among the authors. A.J.R. performed all simulations and wrote the paper. F.V.L. helped with the paper writing and supervised all the technical aspects of the work. R.P.S. and K.A.B. assisted in manuscript preparation and provided technical guidance for the performed work.

Conflicts of Interest: The authors declare no conflict of interest.

References

1. Marbán, G.; Valdés-Solís, T. Towards the hydrogen economy? *Int. J. Hydrogen Energy* **2007**, *32*, 1625–1637.

2. Lima, F.V.; Daoutidis, P.; Tsapatsis, M.; Marano, J.J. Modeling and optimization of membrane reactors for carbon capture in integrated gasification combined cycle units. *Ind. Eng. Chem. Res.* **2012**, *51*, 5480–5489.

3. Woods, M.C.; Capicotto, P.J.; Haslbeck, J.L.; Kuehn, N.J.; Matuszewski, M.; Pinkerton, L.L.; Rutkowski, M.D; Schoff, R.L.; Vaysman, V. *Cost and Performance Baseline for Fossil Energy Plants. Volume 1: Bituminous Coal and Natural Gas to Electricity*; Final Report, Technical Report Revision 1, DOE/NETL-2007/1281. U.S. Department of Energy: Washington, DC, USA, 2007.

4. Lima, F.V.; Daoutidis, P.; Tsapatsis, M. Modeling, optimization, and cost analysis of an IGCC plant with a membrane reactor for carbon capture. *AIChE J.* **2016**.

5. Berchtold, K.A.; Singh, R.P.; Young, J.S.; Dudeck, K.W. Polybenzimidazole composite membranes for high temperature synthesis gas separations. *J. Membr. Sci.* **2012**, *415–416*, 265–270.

6. Singh, R.P.; Berchtold, K.A. Chapter 6—H_2 Selective Membranes for Precombustion Carbon Capture. In *Novel Materials for Carbon Dioxide Mitigation Technology*; Morreale, F.S., Ed.; Elsevier: Amsterdam, The Netherlands, 2015; pp. 177–206.

7. Li, X.; Singh, R.P.; Dudeck, K.W.; Berchtold, K.A.; Benicewicz, B.C. Influence of polybenzimidazole main chain structure on H_2/CO_2 separation at elevated temperatures. *J. Membr. Sci.* **2014**, *461*, 59–68.

8. Zou, J.; Ho, W.S.W. Hydrogen purification for fuel cells by carbon dioxide removal membrane followed by water gas shift reaction. *J. Chem. Eng. Japan* **2007**, *40*, 1011–1020.

9. Huang, J.; El-Azzami, L.; Ho, W.S.W. Modeling of CO_2-selective water-gas-shift membrane reactor for fuel cell. *J. Membr. Sci.* **2005**, *261*, 67–75.

10. Singh, R.P.; Dahe, G.J.; Dudeck, K.W.; Welch, C.F.; Berchtold, K.A. High temperature polybenzimidazole hollow fiber membranes for hydrogen separation and carbon dioxide capture from synthesis gas. *Energy Procedia* **2014**, *63*, 153–159.

11. Merkel, T.C.; Zhou, M.; Baker, R.W. Carbon dioxide capture with membranes at an IGCC power plant. *J. Membr. Sci.* **2012**, *389*, 441–450.

12. Caravella, A.; di Maio, F.P.; di Renzo, A. Optimization of membrane area and catalyst distribution in a permeative-stage membrane reactor for methane steam reforming. *J. Membr. Sci.* **2008**, *321*, 209–221.

13. Caravella, A.; di Maio, F.P.; di Renzo, A. Computation study of staged membrane reactor configurations for methane steam reforming. I. Optimization of stage lengths. *AIChE J.* **2010**, *56*, 248–258.

14. Caravella, A.; di Maio, F.P.; di Renzo, A. Computation study of staged membrane reactor configurations for methane steam reforming. II. Effect of number of stages and catalyst amount. *AIChE J.* **2010**, *56*, 259–267.

15. Adrover, M.E.; López, E.; Borio, D.O.; Pedernera, M.N. Theoretical study of a membrane reactor for the water gas shift reaction under nonisothermal conditions. *AIChE J.* **2009**, *55*, 3206–3213.

16. Choi, Y.; Stenger, H.G. Water gas shift reaction kinetics and reactor modeling for fuel cell grade hydrogen. *J. Power Sources* **2003**, *124*, 432–439.

17. Kärger, J.; Ruthven, D.M. *Diffusion in Zeolites and Other Microporous Solids*; Wiley-Interscience: New York, NY, USA, 1992.

18. Ghosal, K.; Freeman, B.D. Gas separation using polymer membranes: An overview. *Polym. Adv. Technol.* **1994**, *5*, 673–697.

19. Baker, R.W. *Membrane Technology and Applications*; John Wiley & Sons: Hoboken, NJ, USA, 2012; p. 152.

20. Vora, S.; Brickett, L. *DOE/NETL Advanced Carbon Dioxide Capture R & D Program: Technology Update*; U.S. Department of Energy/National Energy Technology Laboratory: Washington, DC, USA, 2013.

21. Marano, J.J. *Integration of H_2 Separation Membranes with CO_2 Capture and Compression*; Report to DoE, Contract No. DE-AC2605NT41816, DOE/NETL-401/113009; U.S. Department of Energy/National Energy Technology Laboratory: Washington, DC, USA, 2009.

Study of *n*-Butyl Acrylate Self-Initiation Reaction Experimentally and via Macroscopic Mechanistic Modeling

Ahmad Arabi Shamsabadi, Nazanin Moghadam, Sriraj Srinivasan, Patrick Corcoran, Michael C. Grady, Andrew M. Rappe and Masoud Soroush

Abstract: This paper presents an experimental study of the self-initiation reaction of *n*-butyl acrylate (*n*-BA) in free-radical polymerization. For the first time, the frequency factor and activation energy of the monomer self-initiation reaction are estimated from measurements of *n*-BA conversion in free-radical homo-polymerization initiated only by the monomer. The estimation was carried out using a macroscopic mechanistic mathematical model of the reactor. In addition to already-known reactions that contribute to the polymerization, the model considers a *n*-BA self-initiation reaction mechanism that is based on our previous electronic-level first-principles theoretical study of the self-initiation reaction. Reaction rate equations are derived using the method of moments. The reaction-rate parameter estimates obtained from conversion measurements agree well with estimates obtained via our purely-theoretical quantum chemical calculations.

Reprinted from *Processes*. Cite as: Shamsabadi, A.A.; Moghadam, N.; Srinivasan, S.; Corcoran, P.; Grady, M.C.; Rappe, A.M.; Soroush, M. Study of *n*-Butyl Acrylate Self-Initiation Reaction Experimentally and via Macroscopic Mechanistic Modeling. *Processes* **2016**, *4*, 15.

1. Introduction

Acrylic polymers are used widely in coatings, as adhesives and functional additives. Increasingly tight limits on volatile organic contents of paints and coatings [1,2] have required the paints and coatings industries to decrease the level of solvents in their products. To ensure adequate brushability and sprayability of the low (20–40 wt%) solvent-content paint and coatings, polymers with low average molecular weights ($\overline{M}_w < 10,000$) have been produced via high-temperature (120–190 °C) polymerization processes [3–5]. At these high temperatures, secondary reactions such as monomer self-initiation, β-scission, inter/intra-molecular chain transfer and backbiting reactions are influential and thus require study [6].

Thermal polymerization of alkyl acrylates in the absence of external initiators has been reported [7]. The occurrence of monomer self-initiation reaction allows one to use less thermal initiators, typically organic peroxides or azonitriles, which are relatively expensive and as residues can cause defects in the final product, especially on weathering [8]. Studies using electron spray ionization-Fourier transform mass spectrometry (ESI-FTMS), nuclear magnetic resonance (NMR) spectroscopy, and macroscopic mechanistic modeling did not identify the initiating species or the mechanism of initiation in the spontaneous thermal polymerization [7,9]. However, quantum chemical calculations [10–13] together with matrix-assisted laser desorption ionization (MALDI) [14] showed that monomer self-initiation is one of the likely mechanisms of initiation in spontaneous thermal polymerization of alkyl acrylates.

Polymer characterization studies using spectroscopic methods have been carried out to explore the dominant polymerization reactions [7,9]. Pulsed-laser polymerization (PLP) and size exclusion chromatography have been used to study intra-molecular chain transfer to polymer (CTP) reactions in polymerization of alkyl acrylates [15,16]. The rate coefficients of the propagation reactions of styrene [17], and methyl methacrylate (MMA) [18], and chain transfer reactions of butyl methacrylate (BMA) [19] at temperatures above 30 °C have been estimated using PLP. Although the propagation rate coefficient of n-butyl acrylate (n-BA) at 70 °C has been calculated using PLP at 500 Hz [20], the presence of backbiting and β-scission reactions has hindered the prediction of alkyl acrylates' propagation rate coefficient at elevated temperatures [21–23]. At temperatures above 30 °C, intra- and inter-molecular CTP reactions in free-radical polymerization of n-BA [24,25] and intra-molecular CTP reactions in 2-ethylhexyl acrylate polymerization [26] have also been studied using NMR spectroscopy. Although these analytical techniques have been very useful in characterizing acrylate polymers, they alone cannot conclusively determine reaction mechanisms or estimate kinetic parameters of reactions. Macroscopic mechanistic modeling combined with adequate polymer sample measurements has proven to be a powerful tool to estimate the rate coefficients of individual reactions. Macroscopic mechanistic models have also been used extensively to estimate the rate coefficients of initiation by conventional thermal initiators, propagation, chain transfer and termination reactions in free-radical polymerization of acrylates [9,16,27–32]. Kinetic parameters of several reactions in spontaneous thermal polymerization of n-BA under seemingly oxygen-free conditions (solvent was bubbled with nitrogen but not n-BA, and a nitrogen blanket was used inside the reactor) were estimated through detailed macroscopic mechanistic modeling, and the entire initiation was attributed to monomer self-initiation, leading to an unrealistically-large self-initiation rate coefficient [9]. Styrene and MMA self-initiation apparent rate coefficients estimated through macroscopic mechanistic modeling have been reported [27,28]. A macroscopic mechanistic model of n-BA spontaneous polymerization, which accounted for initiation only by the monomer, was used to estimate the n-BA self-initiation rate coefficient from measurements of conversion under seemingly oxygen-free conditions [9], again leading to a very large reaction rate coefficient for monomer self-initiation, because the entire monomer conversion was attributed to initiation only by the monomer [29].

Macroscopic mechanistic polymerization models are more useful than semi-empirical models. The accuracy of mechanistic polymerization models strongly depends on our quantitative understanding of individual reactions occurring during the course of polymerization. These models have been used to study low and high-temperature polymerization reactions of n-BA [33,34]. The method of moments [35–37] or the "tendency modeling" approach [38–41] can be used to derive rate equationsneeded in macroscopic mechanistic models. Models obtained using the tendency-modeling rate equations do not account for the number of monomer units in polymer chains, and therefore they are less complex than models that are based on the method-of-moments rate equations. Models that are based on the method of moments can be used to estimate kinetic rate coefficients more accurately. Applications of both types of models can be found in the literature [42–46].

In our previous work, we studied self-initiation of alkyl acrylates such as methyl, ethyl and n-butyl acrylate as well as methyl methacrylate theoretically using density functional theory [10–13]. We found that self-initiation of alkyl acrylates has three elementary reaction steps in series:

(i) Two monomers react and form a singlet diradical

$$M + M \quad \rightarrow \quad {}^*MM^*_s \tag{1}$$

(ii) The singlet diradical then undergoes intersystem crossing to form a triplet diradical:

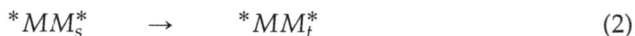

$$ {}^*MM^*_s \quad \rightarrow \quad {}^*MM^*_t \tag{2}$$

(iii) The triplet diradical finally reacts with a third monomer, leading to the formation of two mono-radicals:

$$ {}^*MM^*_t + M \quad \rightarrow \quad MM^* + M^* \tag{3}$$

We also calculated the kinetic parameters of the three preceding reactions [10–12]. The first reaction is the fastest reaction, whereas the second reaction is the slowest (rate limiting) reaction. Therefore, the overall (apparent) self-initiation reaction:

$$3M \quad \xrightarrow{k_i} \quad MM^* + M^* \tag{4}$$

is second order.

In this paper, we experimentally estimate the kinetic parameters (activation energy and frequency factor) of the overall (apparent) n-BA self-initiation reaction in Equation (4) from monomer conversion measurements using a macroscopic mechanistic polymerization reactor model guided by our first-principles investigations of the mechanism. These estimates are compared with our existing estimates of the same parameters obtained via quantum chemical calculations.

The organization of the rest of this paper is as follows. Section 2 presents the mathematical model. Section 3 discusses the experimental and analytical procedures. Section 4 describes the parameter estimation study. Finally, concluding remarks are given in Section 5.

2. Mathematical Modeling

2.1. Reaction Mechanisms

The most likely polymerization reactions occurring in spontaneous thermal solution polymerization of n-BA in the absence of oxygen are given in Table 1. The reactions include monomer self-initiation [10–12], secondary and tertiary chain propagation, intra-molecular chain transfer to polymer (backbiting), inter-molecular chain transfer to polymer, β-scission, chain transfer to monomer [47], chain transfer to solvent, termination by combination, and termination by disproportionation reactions [48,49]. While not all of these reactions strongly affect monomer conversion, the list of the reactions is given here for completeness. For example, no solvent has been used in the study presented herein. Here, M and S denote the monomer and solvent, respectively. U_n represents a dead polymer chain with n monomer units and a terminal double bond. D_n is a dead polymer chain with n monomer units but without a terminal double bond. $R_n{}^*$ represents a secondary radical with n monomer units. $R_n{}^{**}$ is a tertiary radical with n monomer units generated through intermolecular CTP reactions. $R_n{}^{***}$ denotes a tertiary radical with n monomer units formed by backbiting reactions. SCB and LCB represent a short and a long chain branching point, respectively.

Inter- and intra-molecular CTP reactions lead to the formation of tertiary radicals, which are capable of undergoing propagation [50] and β-scission reactions. The β-scission reactions produce secondary radicals and dead polymer chains with a terminal double bond. This led to the generation of shorter dead polymer chains, thus lowering the average molecular weight of the polymer product.

Table 1. Polymerization reactions [34].

a. Apparent monomer self-initiation reaction

$$3M \xrightarrow{k_i} R_1^* + R_2^*$$

b. Propagation reactions

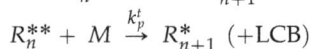

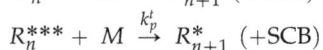

$$R_n^* + M \xrightarrow{k_p} R_{n+1}^*$$

$$R_n^{**} + M \xrightarrow{k_p^t} R_{n+1}^* \ (+LCB)$$

$$R_n^{***} + M \xrightarrow{k_p^t} R_{n+1}^* \ (+SCB)$$

$$R_n^* + U_m \xrightarrow{k_{mac}} R_{n+m}^{**}$$

c. Backbiting reactions ($n > 2$)

$$R_n^* \xrightarrow{k_{bb}} R_n^{***}$$

d. β-scission reactions ($n > 3$)

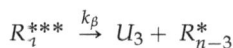

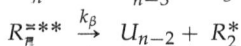

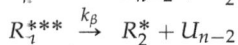

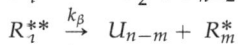

$$R_n^{***} \xrightarrow{k_\beta} U_3 + R_{n-3}^*$$

$$R_n^{***} \xrightarrow{k_\beta} R_{n-3}^* + U_3$$

$$R_n^{***} \xrightarrow{k_\beta} U_{n-2} + R_2^*$$

$$R_n^{***} \xrightarrow{k_\beta} R_2^* + U_{n-2}$$

$$R_n^{**} \xrightarrow{k_\beta} U_{n-m} + R_m^*$$

$$R_n^{**} \xrightarrow{k_\beta} R_m^* + U_{n-m}$$

e. Intermolecular chain transfer to polymer reactions

$$R_n^* - D_m \xrightarrow{mk_{trP}} D_n + R_m^{**}$$

$$R_n^* - U_m \xrightarrow{mk_{trP}} D_n + R_m^{**}$$

f. Chain transfer to monomer reactions

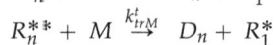

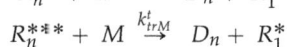

$$R_n^* + M \xrightarrow{k_{trM}} D_n + R_1^*$$

$$R_n^{**} + M \xrightarrow{k_{trM}^t} D_n + R_1^*$$

$$R_n^{***} + M \xrightarrow{k_{trM}^t} D_n + R_1^*$$

g. Chain transfer to solvent reactions

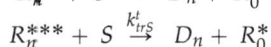

$$R_n^* + S \xrightarrow{k_{trS}} D_n + R_0^*$$

$$R_n^{**} + S \xrightarrow{k_{trS}^t} D_n + R_0^*$$

$$R_n^{***} + S \xrightarrow{k_{trS}^t} D_n + R_0^*$$

Table 1. *Cont.*

h. Termination by coupling reactions

$$R_n^* + R_m^* \xrightarrow{k_{tc}} D_{n+m}$$
$$R_n^* + R_m^{**} \xrightarrow{2k_{tc}^t} D_{n+m}$$
$$R_n^* + R_m^{***} \xrightarrow{2k_{tc}^t} D_{n+m}$$
$$R_n^{**} + R_m^{**} \xrightarrow{k_{tc}^{tt}} D_{n+m}$$
$$R_n^{**} + R_m^{***} \xrightarrow{2k_{tc}^{tt}} D_{n+m}$$
$$R_n^{***} + R_m^{***} \xrightarrow{k_{tc}^{tt}} D_{n+m}$$

i. Termination by disproportionation reactions

$$R_n^* + R_m^* \xrightarrow{k_{td}} D_n + U_m$$
$$R_n^* + R_m^{**} \xrightarrow{k_{td}^t} D_n + U_m$$
$$R_n^* + R_m^{**} \xrightarrow{k_{td}^t} D_m + U_n$$
$$R_n^* + R_m^{***} \xrightarrow{k_{td}^t} D_n + U_m$$
$$R_n^* + R_m^{***} \xrightarrow{k_{td}^t} D_m + U_n$$
$$R_n^{**} + R_m^{**} \xrightarrow{k_{td}^{tt}} D_n + U_m$$
$$R_n^{**} + R_m^{***} \xrightarrow{k_{td}^{tt}} D_n + U_m$$
$$R_n^{**} + R_m^{***} \xrightarrow{k_{td}^{tt}} D_m + U_n$$
$$R_n^{***} + R_m^{***} \xrightarrow{k_{td}^{tt}} D_n + U_m$$

2.2. Rate Equations

Reaction rate equations are derived using the method of moments. We assume that all reactions given in Table 1 except for the self-initiation reaction are elementary. As expected, accounting for β-scission and inter-molecular CTP reactions leads to closure problems; that is, a moment of a chain length distribution depends on a higher moment of the same or different distributions [36]. For example, as the inter-molecular CTP reaction rate coefficient depends on the number of polymerized monomer units in the dead polymer chains, the zeroth moments of the chain length distributions of the dead polymer chains depend on their first moments. To address this problem, for each chain length distribution, a specific distribution model is assumed to derive an approximation that relates the third moment of the distribution to lower moments of the same distribution. In particular, chain length distributions of the dead polymer chains with or without a terminal double bond are assumed to be re-scaled Gamma distributions [51], and the chain length distribution of the tertiary radicals R_n^{**} is assumed to have a normal distribution.

The resulting rate equations; that is, the production rate equations for M, S, R_0^*, R_1^*, R_2^*, R_3^*, and the zeroth, first, and second moments of dead polymer, secondary radical, and tertiary radical chain length distributions, are given in the Appendix. $[X]$ represents the molar concentration of species X. δ_0^*, δ_1^*, and δ_2^* are the zeroth, first, and second moments of the secondary radical chain length distributions. δ_0^{**}, δ_1^{**}, δ_2^{**}, and δ_3^{**} are the zeroth, first, second, and third moments of the chain length distribution of the tertiary radicals generated by the intermolecular CTP reactions. δ_0^{***}, δ_1^{***}, and δ_2^{***} are the zeroth, first, and second moments of the chain length distribution of the tertiary radicals formed by the backbiting reactions. The jth moments of the chain length distributions of the live and dead polymer chains are:

$$\delta_j^* = \sum_{n=0}^{\infty} n^j [R_n^*], \quad \delta_j^{**} = \sum_{n=1}^{\infty} n^j [R_n^{**}], \quad \delta_j^{***} = \sum_{n=1}^{\infty} n^j [R_n^{***}]$$

$$\lambda_j = \sum_{n=1}^{\infty} n^j [D_n], \quad \Gamma_j = \sum_{n=1}^{\infty} n^j [U_n]$$

and $\delta_0 = \delta_0^* + \delta_0^{**} + \delta_0^{***}$.

2.3. Batch Reactor Model

Mole balances on all species and balances on the moments lead to a batch reactor model that consists of 21 first-order ordinary differential equations:

$$\frac{d[J]}{dt} = r_J, \quad [J](0) = [J]_0$$

where $J = M, S, R_0^*, R_1^*, R_2^*, R_3^*, \delta_0^*, \delta_1^*, \delta_2^*, \delta_0^{**}, \delta_1^{**}, \delta_2^{**}, \delta_0^{***}, \delta_1^{***}, \delta_2^{***}, \lambda_0, \lambda_1, \lambda_2, \Gamma_0, \Gamma_1, \Gamma_2$. All initial concentrations are assigned to be zero except for that of monomer, which is nonzero and is denoted by $[M]_0$. The monomer conversion is calculated using:

$$X = \frac{[M]_0 - [M]}{[M]_0}$$

In this model, volume effects and diffusional limitations are ignored as most of conversion measurements reported herein are below 50%. Also, depropagation is not considered, as it is insignificant for alkyl acrylate monomers. However, it is significant for methacrylate monomers [5].

3. Experimental and Analytical Procedures

The monomer, 98% n-butyl acrylate stabilized with 50 ppm of 4-methoxyphenol as inhibitor, is from Alfa Aesar. Batch reactors are 4-inch stainless steel Swagelok tubes (Swagelok Inc., Huntingdon Valley, PA, USA), capped at both ends with

Swagelok stainless steel caps. The 4-inch length gives a reaction volume of 4.8 mL. These tubes can withstand pressures up to 3300 psig.

For each set of batch experiments, 30–50 mL of n-BA is dripped through an inhibitor removal column DHR-4, from Scientific Polymer Products of Ontario, New York, in order to remove the inhibitor. The inhibitor-free monomer is collected in a 50-mL round-bottom flask equipped with a standard taper 24/40 ground glass joint. After one hour of UHP nitrogen bubbling, we remove the needles and wrap the rubber septum tightly with aluminum foil secured with tight rubber bands. The round-bottom flask with the inhibitor-free, nitrogen-bubbled n-BA, the open reaction tubes, and the tube caps are then moved to the vacuum-nitrogen-purge chamber of a nitrogen-atmosphere glove box (LC Technology Solutions, Salisbury, USA). After several vacuum-nitrogen purge cycles, the flask, tubes and caps are then moved to the main chamber of the glove box, in which oxygen is removed reactively and water physically from the nitrogen gas inside. The oxygen concentration in the glove box is kept below 0.1 ppm. Inside the glove box, the sealed flask is then opened and 2.5 mL of monomer are pipetted into each reaction tube, after which the other end cap is attached and tightened. Upon removal from the glove box, each reaction tube is weighed, and the weight is recorded. The fluidized sand bath is then heated to a desired constant reaction temperature. After maintaining the sand bath at the desired temperature for several hours, two reactor tubes at a time are placed in the sand bath and are then pulled out of the fluidized sand bath after a specific period of time. The tubes are then cooled quickly in a cold water bath to stop further polymerization. After drying each tube, the reaction tube is weighed, and the weight is recorded. If any tube shows a weight loss, the tube is discarded from the experiment. The content of each tube is then emptied into vials. The time that a reactor tube spends in the sand bath minus one minute is considered as the reaction time of the reactor tube. Our previous studies had shown that it takes approximately one minute for the monomer inside a reactor tube to reach the sand bath temperature.

The conversion in each reaction tube is measured with a gravimetric method. A 57 mm aluminum dish is weighted (wgt1). Then, 1.5 mL of the reaction mass is pipetted from each vial into the aluminum dish, and the dish is then weighed (wgt2). Three milliliters of toluene is then added to the reaction mass, which completely dissolves in the added toluene. Next, the dish is placed in a vacuum oven overnight at 50 °C to allow the solvent and the unreacted monomer to evaporate. The dish is then weighed a third time (wgt3). The % conversion is calculated using the equation $((wgt1 - wgt3) \times 100)/(wgt1 - wgt2)$.

4. Results and Discussion

Parameter Estimation

Rate coefficients of all reactions except for the monomer self-initiation reaction are given in Table 2.

Table 2. Reaction rate coefficient values.

Parameter	Frequency Factor		Activation Energy (kJ.mol^{-1})	Ref.
k_p	2.21×10^7	L·mol^{-1}·s^{-1}	17.9	[30]
k_p^t	1.20×10^6	L·mol^{-1}·s^{-1}	28.6	[49]
k_{bb}	7.41×10^7	s^{-1}	32.7	[33]
k_{trM}	2.90×10^5	L·mol^{-1}·s^{-1}	32.6	[47]
k_t	3.89×10^9	L·mol^{-1}·s^{-1}	8.4	[31]
k_t^{tt}	5.30×10^9	L·mol^{-1}·s^{-1}	19.6	[34]
k_β	1.49×10^9	s^{-1}	63.9	[34]
k_{trP}	4.01×10^3	L·mol^{-1}·s^{-1}	29.0	[15]
C_{trS}	1.07×10^2		35.4	[34]

Table 3. Reaction rate coefficient definitions and correlations [34,48].

$$k_t = k_{tc} + k_{td}$$
$$k_t^t = k_{tc}^t + k_{td}^t$$
$$k_t^{tt} = k_{tc}^{tt} + k_{td}^{tt}$$
$$k_{td} = \delta_s k_t$$
$$k_{td}^{tt} = \delta_t k_t^{tt}$$
$$k_{td}^t = \delta_{st} \sqrt{k_t k_t^{tt}}$$
$$k_{tc} = (1 - \delta_s) k_t$$
$$k_{tc}^{tt} = (1 - \delta_t) k_t^{tt}$$
$$k_{tc}^t = (1 - \delta_{st}) \sqrt{k_t k_t^{tt}}$$
$$k_{trS} = C_{trS} k_p$$
$$k_{trM}^t = \frac{k_p^t}{k_p} k_{trM}$$
$$k_{mac} = \gamma k_p$$

The unknown rate coefficient, k_i, is estimated from monomer-conversion measurements. Reaction rate coefficient definitions, correlations and dimensionless kinetic parameter values are provided in Tables 3 and 4.

Table 4. Dimensionless kinetic parameter values [34,48].

Parameter	Dimensionless Value
δ_s	0.1
δ_{st}	0.7
δ_t	0.9
γ	0.5

The system of ordinary differential equations (ODEs) is integrated numerically using the MATLAB routine bvp5c. For parameter estimation, the MATLAB function fminsearch, which is based on the Nelder-Mead simplex algorithm, is used to find the value of k_i that minimizes the sum of the squared errors between measurements and model-predicted values of conversion.

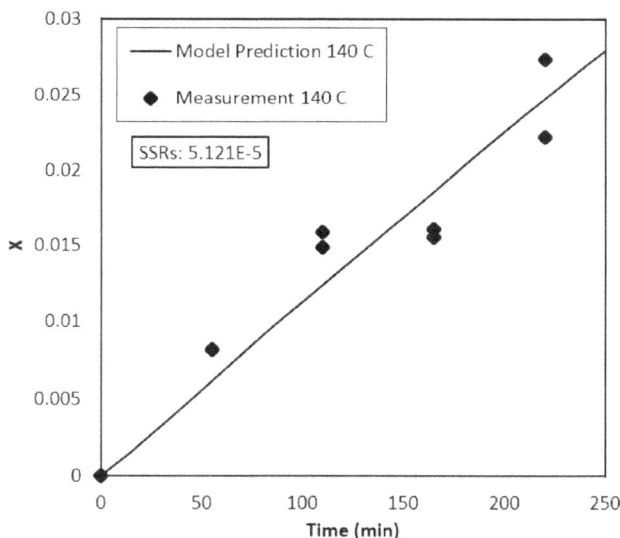

Figure 1. Measurements and model prediction of monomer conversion at 140 °C. SSRs = sum of squared residuals.

Figures 1–5 show measurements and model predictions of n-BA conversion at different constant reactor temperatures (140, 160, 180, 200 and 220 °C). At each temperature, k_i, is estimated by fitting the model predictions to the measurements. As can be seen, the model fits the measurements very well at all four temperatures.

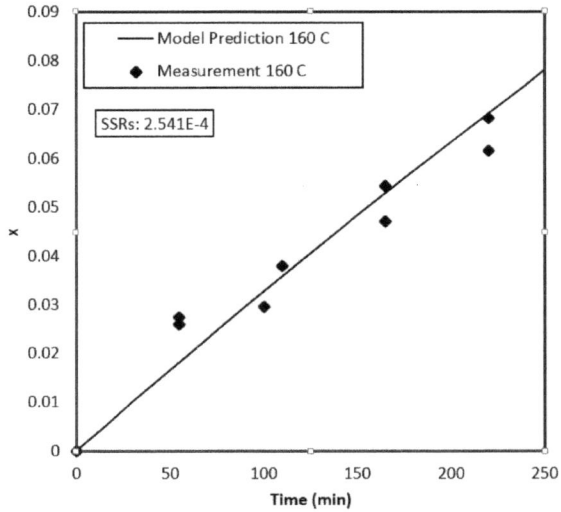

Figure 2. Measurements and model prediction of monomer conversion at 160 °C.

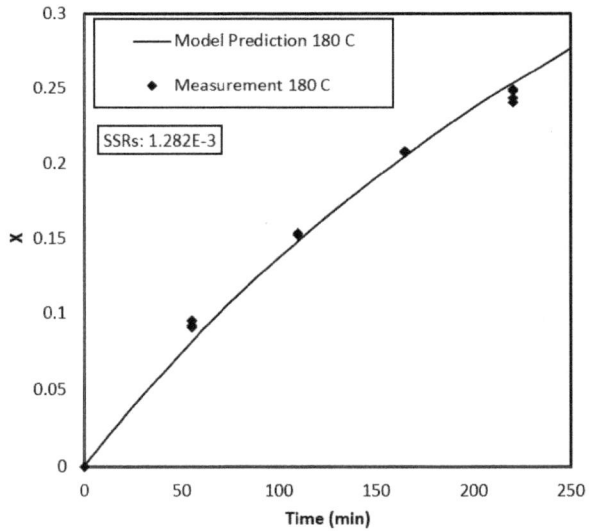

Figure 3. Measurements and model prediction of monomer conversion at 180 °C.

Once the reaction rate coefficient is estimated at 140, 160, 180, 200 and 220 °C, the reaction frequency factor and activation energy are estimated using the Arrhenius plot (see Figure 6).

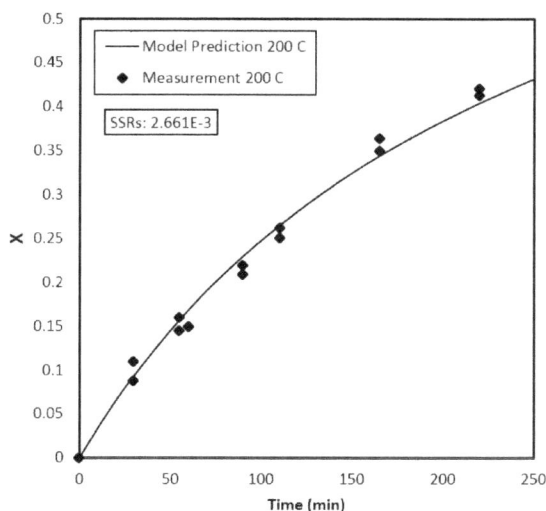

Figure 4. Measurements and model prediction of monomer conversion at 200 °C.

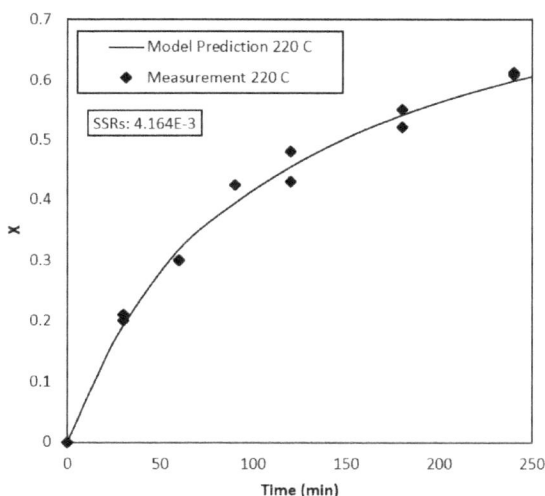

Figure 5. Measurements and model prediction of monomer conversion at 220 °C.

The estimated frequency factor and activation energy as well as their 95% confidence intervals are given in Table **??**. As can be seen in this table, the experimentally estimated self-initiation rate coefficients are in excellent agreement with those calculated using quantum chemical calculations (within a factor of 2–3 of those reported in [12]).

191

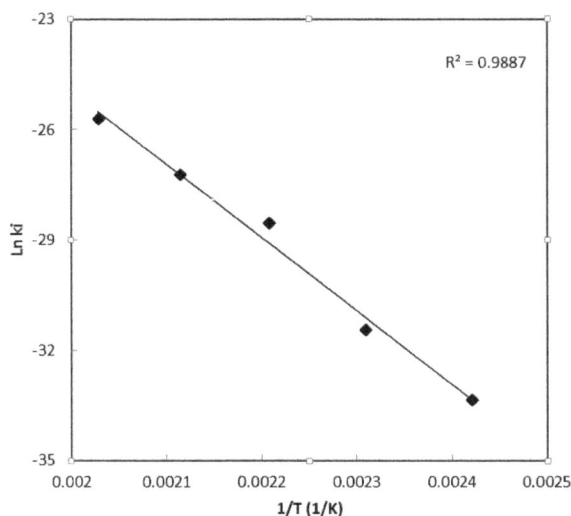

Figure 6. Arrhenius plot of k_i to estimate E_i and A_i .

After 250 min of polymerization at 140 and 160 °C, respectively, the conversion is less than 3% (Figure 1) and less than 8% (Figure 2). Theses indicate that the n-BA self-initiation reaction is slow at 160 °C or a lower temperature. However, as temperature increases (Figures 3–5), the contribution of self-initiation to the polymerization increases (conversion of more than 25% at 180 °C, about 40% at 200 °C, and about 60% at 220 °C, after 250 min of reaction). At 160 °C, conversion increases linearly from 2.5% to 6.5% between 55 and 220 min. At 180 °C, the conversion increases to 25% after 220 min, and the rate of its increase is slightly higher initially. At 200 °C, a 25% conversion is achieved after 110 min. Again, the conversion increases faster initially at 200 °C. At 220 °C, less than 55 min is sufficient to achieve a 30% conversion. Figures 1–5 indicate that at 140–220 °C the conversion does not level off during the 250 min, implying that the polymerization can continue beyond 250 min. Figure 7 shows the high sensitivity of the conversion predictions to the self-initiation rate coefficient, implying the low uncertainty of the estimate of the rate coefficient at each of the temperatures.

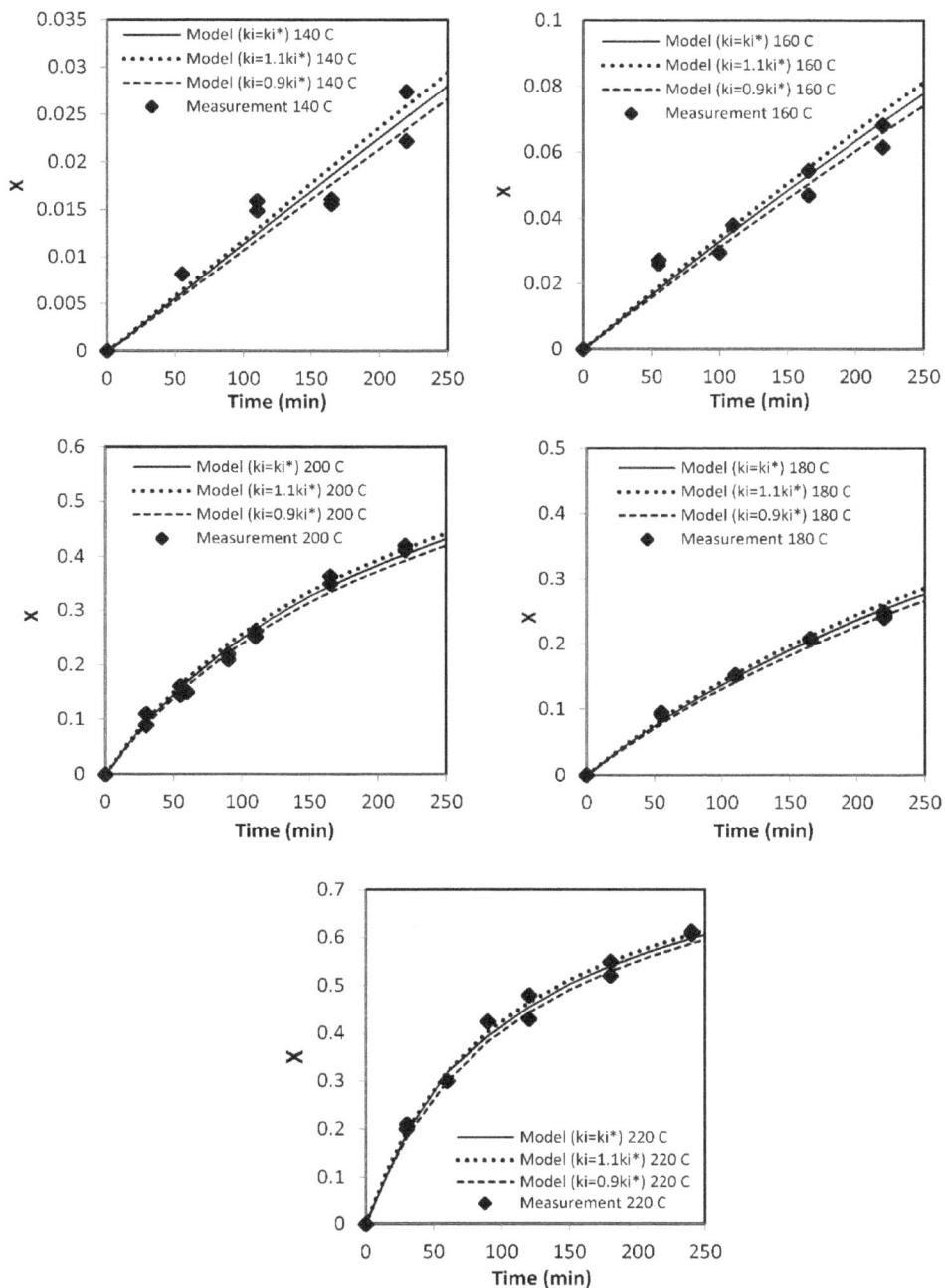

Figure 7. Model prediction and measurements of conversion as well as the sensitivity of predicted conversion to 10% changes in k_i at the five temperatures.

Table 5. Experimentally-estimated and theoretical values of k_i ($M^{-1} \cdot s^{-1}$), E_i ($kJ \cdot mol^{-1}$) and A_i ($M^{-1} \cdot s^{-1}$).

Temperature	This Work	Theoretical [11]	Theoretical [12]
T	k_i	k_i	k_i
413	3.30×10^{-15}	2.81×10^{-18}	1.04×10^{-14}
433	2.20×10^{-14}	2.86×10^{-17}	4.72×10^{-14}
453	4.00×10^{-13}	2.37×10^{-16}	1.95×10^{-13}
473	1.50×10^{-12}	1.64×10^{-15}	7.11×10^{-13}
493	6.80×10^{-12}	9.74×10^{-15}	2.34×10^{-12}

Parameter	This Work	Theoretical [11]	Theoretical [12]
E_i	165.51 ± 4.52	172.50	115.00
$\ln A_i$	14.86 ± 1.20	9.68	1.38

5. Concluding Remarks

An experimental study of the self-initiation reaction of n-BA in free-radical polymerization was presented. The frequency factor and activation energy of the monomer self-initiation reaction were estimated from batch-reactor monomer-conversion measurements from spontaneous polymerization of n-BA in the absence of oxygen, using a macroscopic mechanistic mathematical model of the reactor. A comparison of the estimated monomer self-initiation reaction rate constant with estimates obtained via quantum chemical calculations showed satisfactory agreement. These experimental estimates quantify, for the first time, the sole contribution of the monomer self-initiation to the initiation of n-BA polymerization.

When two alkyl acrylate monomers react, they form a diradical that undergoes spin transition and subsequently reacts with a third monomer and forms two monoradicals, which initiate polymerization [10–12]. The rate coefficients of alkyl (methyl, ethyl, n-butyl) acrylate monomers have similar values [10–12]. However, when two methacylate monomers react, they form a singlet diradical that can undergo three major competing parallel reactions (one generates a dimer and the other two lead to the formation of monoradicals, which initiate polymerization) [13]. Because of the competing parallel reaction that generates a dimer, methacylate monomers have slower monomer-initiated polymerization in comparison with alkyl acrylate monomers.

Acknowledgments: This material is based upon work partially supported by the National Science Foundation under Grant CBET-1160169. A.M.R. acknowledges support from the NSF under grant CBET-1159736. Any opinions, findings, and conclusions or recommendations expressed in this material are those of the authors and do not necessarily reflect the views of the National Science Foundation.

Author Contributions: M.S., A.M.R. and M.C.G. conceived and designed the experiments. A.A.S. and P.C. performed the experiments. N.M. and M.S. derived the reaction rate equations. N.M. and A.A.S. conducted the numerical simulations. S.S., A.A.S., and M.S. analyzed the data. A.A.S., N.M., S.S., A.M.R. and M.S. wrote the paper.

Conflicts of Interest: The authors declare no conflict of interest.

Appendix: Reaction Rate Equations

$$r_M = -3k_i[M]^2 - k_p\,[M]\,\delta_0^* \;-\; k_p^t\,[M]\,(\delta_0^{**} + \delta_0^{***}) \;-\; k_{trM}\,[M]\,\delta_0^* - k_{trM}^t\,[M]\,(\delta_0^{**} + \delta_0^{***})$$

$$r_S = -k_{trS}\,[S]\,\delta_0^* - k_{trS}^t\,[S]\,(\delta_0^{**} + \delta_0^{***})$$

$$r_{R_0^*} = -k_p\,[M]\,[R_0^*] - k_{mac}[R_0^*]\Gamma_0 - k_{tc}[R_0^*]\delta_0^* - k_{tc}^t\,[R_0^*]\,(\delta_0^{**} + \delta_0^{***})$$

$$r_{R_1^*} = k_i[M]^2 + k_p\,[M]\,([R_0^*] - [R_1^*]) - k_{mac}[R_1^*]\Gamma_0 + 2k_\beta\delta_0^{**}$$
$$+k_{trm}\,[M]\,\delta_0^* + k_{trm}^t\,[M]\,(\delta_0^{**} + \delta_0^{***}) - (k_{tc} + 2k_{td})\,[R_1^*]\,\delta_0^*$$
$$- (k_{tc}^t + 2k_{td}^t)\,[R_1^*]\,(\delta_0^{**} + \delta_0^{***}) - k_{trp}^t\,[R_1^*]\,(\lambda_1 + \Gamma_1)$$

$$r_{R_2^*} = k_i[M]^2 + k_p\,[M]\,([R_1^*] - [R_2^*]) - k_{mac}[R_2^*]\,\Gamma_0 + 2k_\beta\,(\delta_0^{***} + 2\delta_0^{**})$$
$$- k_{trP}\,(\lambda_1 + \Gamma_1)\,[R_2^*] - (k_{trM}\,[M])\,[R_2^*] - (k_{tc} + 2k_{td})\,[R_2^*]\,\delta_0^*$$
$$- (k_{tc}^t + 2k_{td}^t)\,[R_2^*]\,(\delta_0^{**} + \delta_0^{***})$$

$$r_{R_3^*} = k_p\,[M]([R_2^*] - [R_3^*]) + k_p^t\,[M]\,[R_2^{**}] - k_{mac}[R_3^*]\Gamma_0 - k_{bb}[R_3^*] + 2k_\beta\delta_0^{**}$$
$$- k_{trP}\,(\lambda_1 + \Gamma_1)\,[R_3^*] - (k_{trM}\,[M])\,[R_3^*] - (k_{tc} + 2k_{td})\,[R_3^*]\delta_0^*$$
$$- (k_{tc}^t + 2k_{td}^t)\,[R_3^*]\,(\delta_0^{**} + \delta_0^{***})$$

$$r_{\delta_0^*} = 2k_i[M]^2 - k_{mac}\Gamma_0\delta_0^* - k_{bb}\delta_0^* + 2k_\beta\,(\delta_1^{**} - 3\delta_0^{**} + 2\delta_0^{***}) - k_{trP}\,(\lambda_1 + \Gamma_1)\,\delta_0^*$$
$$+ (k_p^t\,[M] + k_{trM}^t\,[M])\,(\delta_0^{**} + \delta_0^{***}) - (k_{tc} + k_{td})\,\delta_0^{*2}$$
$$-2\,(k_{tc}^t + k_{td}^t)\,\delta_0^*\,(\delta_0^{**} + \delta_0^{***})$$

$$r_{\delta_1^*} = 3k_i[M]^2 + k_p\,[M]\,\delta_0^* + k_p^t\,[M]\,(\delta_0^{**} + \delta_1^{**} + \delta_0^{***} + \delta_1^{***}) - k_{mac}\Gamma_0\delta_1^* - k_{bb}\delta_1^*$$
$$+k_\beta\,(2\delta_1^{***} - 2\delta_0^{***} + \delta_2^{**} - 5\delta_1^{**} + 6\delta_0^{**}) - k_{trP}\,(\lambda_1 + \Gamma_1)\,\delta_1^*$$
$$+k_{trM}\,[M]\,(\delta_0^* - \delta_1^*) + k_{trM}^t\,[M]\,(\delta_0^{**} + \delta_0^{***}) - (k_{tc} + k_{td})\,\delta_0^*\delta_1^*$$
$$-2\,(k_{tc}^t + k_{td}^t)\,\delta_1^*\,(\delta_0^{**} + \delta_0^{***})$$

$$r_{\delta_2^*} = 5k_i[M]^2 + k_p\,[M]\,(2\delta_1^* + \delta_0^*) + k_p^t\,[M]\,(\delta_2^{**} + 2\delta_1^{**} + \delta_0^{**} + \delta_2^{***} + 2\delta_1^{***} + \delta_0^{***})$$

$$-k_{mac}\Gamma_0\delta_2^* - k_{bb}\delta_2^* - k_{trP}\,(\lambda_1 + \Gamma_1)\,\delta_2^* + k_{trM}\,[M]\,(\delta_0^* - \delta_2^*)$$

$$+k_{trM}^t\,[M]\,(\delta_0^{**} + \delta_0^{***}) - (k_{tc} + k_{td})\,\delta_0^*\delta_2^*$$

$$-2\left(k_{tc}^t + k_{td}^t\right)\delta_2^*\left(\delta_0^{**} + \delta_0^{***}\right)$$

$$+k_\beta\left(2\delta_2^{***} - 12\delta_1^{***} + 26\delta_0^{***} + \tfrac{2}{3}\delta_3^{**} - 5\delta_2^{**} + \tfrac{37}{3}\delta_1^{**} - 10\delta_0^{**}\right)$$

$$r_{\delta_0^{**}} = -k_p^t\,[M]\,\delta_0^{**} + k_{mac}\Gamma_0\delta_0^* - 2k_\beta\left(\delta_1^{**} - 3\delta_0^{**}\right) + k_{trP}\,(\lambda_1 + \Gamma_1)\,\delta_0^* - \left(k_{trM}^t\,[M]\right)\delta_0^{**}$$

$$-2\left(k_{tc}^t + k_{td}^t\right)\delta_0^*\delta_0^{**} - \left(k_{tc}^{tt} + k_{td}^{tt}\right)\delta_0^{**}\left(\delta_0^{**} + 2\delta_0^{***}\right)$$

$$r_{\delta_1^{**}} = -k_p^t\,[M]\,\delta_1^{**} + k_{mac}\left(\delta_0^*\Gamma_1 + \delta_1^*\Gamma_0\right) - 2k_\beta\left(\delta_2^{**} - 3\delta_1^{**}\right) + k_{trP}\delta_0^*\,(\lambda_2 + \Gamma_2)$$

$$-\left(k_{trM}^t\,[M]\right)\delta_1^{**} - 2\left(k_{tc}^t + k_{td}^t\right)\delta_0^*\delta_1^{**} - \left(k_{tc}^{tt} + k_{td}^{tt}\right)\delta_1^{**}\left(\delta_0^{**} + 2\delta_0^{***}\right)$$

$$r_{\delta_2^{**}} = -k_p^t\,[M]\,\delta_2^{**} + k_{mac}\left(\delta_0^*\Gamma_2 + 2\delta_1^*\Gamma_1 + \delta_2^*\Gamma_0\right) - 2k_\beta\left(\delta_3^{**} - 3\delta_2^{**}\right) + k_{trP}\delta_0^*$$

$$(\Gamma_3 + \lambda_3) - \left(k_{trM}^t\,[M]\right)\delta_2^{**} - 2\left(k_{tc}^t + k_{td}^t\right)\delta_0^*\delta_2^{**} - \left(k_{tc}^{tt} + k_{td}^{tt}\right)\delta_2^{**}\left(\delta_0^{**} + 2\delta_0^{***}\right)$$

$$r_{\delta_0^{***}} = -k_p^t\,[M]\,\delta_0^{***} + k_{bb}\delta_0^* - 4k_\beta\delta_0^{***} - \left(k_{trM}^t\,[M]\right)\delta_0^{***} - 2\left(k_{tc}^t + k_{td}^t\right)\delta_0^*\delta_0^{***}$$

$$-\left(k_{tc}^{tt} + k_{td}^{tt}\right)\delta_0^{***}\left(2\delta_0^{**} + \delta_0^{***}\right)$$

$$r_{\delta_1^{***}} = -k_p^t\,[M]\,\delta_1^{***} + k_{bb}\delta_1^* - 4k_\beta\delta_1^{***} - \left(k_{trM}^t\,[M]\right)\delta_1^{***} - 2\left(k_{tc}^t + k_{td}^t\right)\delta_0^*\delta_1^{***}$$

$$-\left(k_{tc}^{tt} + k_{td}^{tt}\right)\delta_1^{***}\left(2\delta_0^{**} + \delta_0^{***}\right)$$

$$r_{\delta_2^{***}} = -k_p^t\,[M]\,[\delta_2^{***}] + k_{bb}\delta_2^* - 4k_\beta\delta_2^{***} - \left(k_{trM}^t\,[M]\right)\delta_2^{***} - 2\left(k_{tc}^t + k_{td}^t\right)\delta_0^*\delta_2^{***}$$

$$-\left(k_{tc}^{tt} + k_{td}^{tt}\right)\delta_2^{***}\left(2\delta_0^{**} + \delta_0^{***}\right)$$

$$r_{\lambda_0} = k_{trP}\delta_0^*\Gamma_1 + (k_{trM}\,[M])\,\delta_0^* + \left(k_{trM}^t\,[M]\right)(\delta_0^{**} + \delta_0^{***}) + 0.5\,(k_{tc} + k_{td})\,\delta_0^*\delta_0^*$$

$$+2\left(k_{tc}^t + k_{td}^t\right)\delta_0^*\left(\delta_0^{**} + \delta_0^{***}\right)$$

$$+0.5\left(k_{tc}^{tt} + k_{td}^{tt}\right)\left(\delta_0^{**}\delta_0^{**} + \delta_0^{***}\delta_0^{***} + 4\delta_0^{**}\delta_0^{***}\right)$$

$$r_{\lambda_1} = k_{trP}\left(\delta_1^*\,(\Gamma_1 + \lambda_1) - \delta_0^*\lambda_2\right) + (k_{trM}\,[M])\,\delta_1^* + \left(k_{trM}^t\,[M]\right)\left(\delta_1^{**} + \delta_1^{***}\right)$$

$$+0.5\,(2k_{tc} + k_{td})\,\delta_0^*\delta_1^*$$

$$+\left(2k_{tc}^t + k_{td}^t\right)\left(\delta_1^*\delta_0^{**} + \delta_1^{**}\delta_0^* + \delta_1^*\delta_0^{***} + \delta_1^{***}\delta_0^*\right)$$

$$+\left(2k_{tc}^{tt} + k_{td}^{tt}\right)\left(0.5\delta_0^{**}\delta_1^{**} + 0.5\delta_0^{***}\delta_1^{***} + \delta_1^{**}\delta_0^{***} + \delta_0^{**}\delta_1^{***}\right)$$

$$r_{\lambda_2} = k_{trP}\left(\delta_2^*\,(\Gamma_1 + \lambda_1) - \delta_0^*\lambda_3\right) + (k_{trM}\,[M])\,\delta_2^* + \left(k_{trM}^t\,[M]\right)\left(\delta_2^{**} + \delta_2^{***}\right) + k_{tc}\left(\delta_0^*\delta_2^* + \delta_1^*\delta_1^*\right)$$

$$+2k_{tc}^t\left(\delta_2^*\left(\delta_0^{**} + \delta_0^{***}\right) + 2\delta_1^*\left(\delta_1^{**} + \delta_1^{***}\right) + \delta_0^*\left(\delta_2^{**} + \delta_2^{**}\right)\right)$$

$$+k_{tc}^{tt}\left(\delta_2^{**}\delta_0^{**} + \delta_1^{**}\delta_1^{**} + \delta_2^{***}\delta_0^{***} + \delta_1^{***}\delta_1^{***} + 2\delta_0^{***}\delta_2^{**} + 2\delta_2^{***}\delta_0^{**}\right.$$

$$\left.+4\delta_1^{***}\delta_1^{**}\right) + 0.5k_{td}\delta_0^*\delta_2^* + k_{td}^t\left(\delta_0^{**}\delta_2^* + \delta_0^*\delta_2^{**} + \delta_0^{***}\delta_2^* + \delta_2^{***}\delta_0^*\right)$$

$$+k_{td}^{tt}\left(0.5\delta_0^{**}\delta_2^{**} + 0.5\delta_0^{***}\delta_2^{***} + \delta_2^{**}\delta_0^{***} + \delta_2^{***}\delta_0^{**}\right)$$

$$r_{\Gamma_0} = -k_{mac}\delta_0^*\Gamma_0 + 2k_\beta \left(2\delta_0^{***} + \delta_1^{**} - 3\delta_0^{**}\right) - k_{trP}\delta_0^*\Gamma_1 + 0.5k_{td}\delta_0^{*2}$$
$$+ 2k_{td}^t\delta_0^* \left(\delta_0^{**} + \delta_0^{***}\right) + 0.5k_{td}^{tt} \left(\delta_0^{**}\delta_0^{**} + 4\delta_0^{**}\delta_0^{***} + \delta_0^{***}\delta_0^{***}\right)$$

$$r_{\Gamma_1} = -k_{mac}\,\delta_0^*\Gamma_1 + k_\beta \left(2\delta_1^{***} + 2\delta_0^{***} + \delta_2^{**} - \delta_1^{**} - 6\delta_0^{**}\right) - k_{trP}\,\delta_0^*\Gamma_2 + 0.5k_{td}\delta_0^*\delta_1^*$$
$$+ k_{td}^t \left(\delta_1^*\delta_0^{**} + \delta_1^{**}\delta_0^* + \delta_0^{***}\delta_1^* + \delta_0^*\delta_1^{**}\right)$$
$$+ 0.5k_{td}^{tt} \left(\delta_0^{**}\delta_1^{**} + \delta_0^{***}\delta_1^{***} + 2\delta_1^{**}\delta_0^{***} + 2\delta_0^{**}\delta_1^{***}\right)$$

$$r_{\Gamma_2} = -k_{mac}\,\delta_0^*\Gamma_2$$
$$+ k_\beta \left(2\delta_2^{***} - 8\delta_1^{***} + 26\delta_0^{***} + \tfrac{2}{3}\delta_3^{**} - \delta_2^{**} + \tfrac{1}{3}\delta_1^{**}\right.$$
$$\left. - 10\delta_0^{**}\right) - k_{trP}\,\delta_0^*\Gamma_3 + 0.5\,k_{td}\,\delta_0^*\delta_2^*$$
$$+ k_{td}^t \left(\delta_0^* \left(\delta_2^{**} + \delta_2^{***}\right) + \delta_2^* \left(\delta_0^{**} + \delta_0^{***}\right)\right)$$
$$+ k_{td}^{tt} \left(0.5\delta_0^{***}\delta_2^{***} + 0.5\delta_0^{**}\delta_2^{**} + \delta_0^{***}\delta_2^{**} + \delta_0^{**}\delta_2^{***}\right)$$

where (assuming re-scaled Gamma distributions for the dead polymer chains and a normal distribution for the tertiary radicals R_n^{**})

$$\lambda_3 \approx \left(\frac{\lambda_2}{\lambda_0\lambda_1}\right) \left(2\lambda_0\lambda_2 - \lambda_1^2\right)$$

$$\Gamma_3 \approx \left(\frac{\Gamma_2}{\Gamma_0\Gamma_1}\right) \left(2\Gamma_0\Gamma_2 - \Gamma_1^2\right)$$

$$\delta_3^{**} \approx 3\delta_2^{**}\delta_1^{**} - 2\delta_1^{**3}$$

References

1. *Consumer and Commercial Products, Group IV: Control Techniques Guidelines in Lieu of Regulations for Miscellaneous Metal Products Coatings, Plastic Parts Coatings, Auto and Light-Duty Truck Assembly Coatings, Fiberglass Boat Manufacturing Materials, and Miscellaneous Industrial Adhesives*; EPA-HQ-OAR-2008-0411; FRL-8725-9; Environmental Protection Agency: Washington, DC, USA, 2008; Volume 73, No. 195; pp. 58481–58490.

2. *Economic Impact and Regulatory Flexibility Analyses of the Final Architectural Coatings VOC Rule*; EPA-452/R-98-002; United States Environmental Protection Agency, Office of Air Quality Planning and Standards: Washington, DC, USA, 1998.

3. Campbell, J.D.; Teymour, F.; Morbidelli, M. High temperature free radical polymerization. 1. Investigation of continuous styrene polymerization. *Macromolecules* **2003**, *36*, 5491–5501.

4. Wang, W.; Hutchinson, R.A. Recent advances in the study of high-temperature free radical acrylic solution copolymerization. *Macromol. React. Eng.* **2008**, *2*, 199–214.

5. Grady, M.C.; Simonsick, W.J.; Hutchinson, R.A. Studies of higher temperature polymerization of n-butyl methacrylate and n-butyl acrylate. *Macromol. Symp.* **2002**, *182*, 149–168.

6. Soroush, M.; Grady, M.C.; Kalfas, G.A. Free-radical polymerization at higher temperatures: Systems impacts of secondary reactions. *Comput. Chem. Eng.* **2008**, *32*, 2155–2167.

7. Quan, C.L.; Soroush, M.; Grady, M.C.; Hansen, J.E.; Simonsick, W.J. High-temperature homopolymerization of ethyl acrylate and n-butyl acrylate: Polymer characterization. *Macromolecules* **2005**, *38*, 7619–7628.

8. Chiefari, J.; Jeffery, J.; Mayadunne, R.T.A.; Moad, G.; Rizzardo, E.; Thang, S.H. Chain transfer to polymer: A convenient route to macromonomers. *Macromolecules* **1999**, *32*, 7700–7702.

9. Rantow, F.S.; Soroush, M.; Grady, M.C.; Kalfas, G.A. Spontaneous polymerization and chain microstructure evolution in high-temperature solution polymerization of n-butyl acrylate. *Polymer* **2006**, *47*, 1423–1435.

10. Srinivasan, S.; Lee, M.W.; Grady, M.C.; Soroush, M.; Rappe, A.M. Computational study of the self-initiation mechanism in thermal polymerization of methyl acrylate. *J. Phys. Chem. A* **2009**, *113*, 10787–10794.

11. Srinivasan, S.; Lee, M.W.; Grady, M.C.; Soroush, M.; Rappe, A.M. Self-initiation mechanism in spontaneous thermal polymerization of ethyl and n-butyl acrylate: A theoretical study. *J. Phys. Chem. A* **2010**, *114*, 7975–7983.

12. Liu, S.; Srinivasan, S.; Tao, J.; Grady, M.C.; Soroush, M.; Rappe, A.M. Modeling Spin-Forbidden Monomer Self-initiation Reactions in Free-Radical Polymerization of Acrylates and Methacrylates. *J. Phys. Chem. A* **2014**, *118*, 9310–9318.

13. Srinivasan, S.; Lee, M.W.; Grady, M.C.; Soroush, M.; Rappe, A.M. Computational Evidence for Self-Initiation in Spontaneous High-Temperature Polymerization of Methyl Methacrylate. *J. Phys. Chem. A* **2011**, *115*, 1125–1132.

14. Srinivasan, S.; Kalfas, G.; Petkovska, V.I.; Bruni, C.; Grady, M.C.; Soroush, M. Experimental study of the spontaneous thermal homopolymerization of methyl and n-butyl acrylate. *J. Appl. Polym. Sci.* **2010**, *118*, 1898–1909.

15. Arzamendi, G.; Plessis, C.; Leiza, J.R.; Asua, J.M. Effect of the intramolecular chain transfer to polymer on PLP/SEC experiments of alkyl acrylates. *Macromol. Theory Simul.* **2003**, *12*, 315–324.

16. Nikitin, A.N.; Hutchinson, R.A.; Buback, M.; Hesse, P. Determination of intramolecular chain transfer and midchain radical propagation rate coefficients for butyl acrylate by pulsed laser polymerization. *Macromolecules* **2007**, *40*, 8631–8641.

17. Buback, M.; Gilbert, R.G.; Hutchinson, R.A.; Klumperman, B.; Kuchta, F.D.; Manders, B.G.; Odriscoll, K.F.; Russell, G.T.; Schweer, J. Critically evaluated rate coefficients for free-radical polymerization. 1. Propagation rate coefficient for styrene. *Macromol. Chem. Phys.* **1995**, *196*, 3267–3280.

18. Beuermann, S.; Buback, M.; Davis, T.P.; Gilbert, R.G.; Hutchinson, R.A.; Olaj, O.F.; Russell, G.T.; Schweer, J.; vanHerk, A.M. Critically evaluated rate coefficients for free-radical polymerization. 2. Propagation rate coefficients for methyl methacrylate. *Macromol. Chem. Phys.* **1997**, *198*, 1545–1560.

19. Nikitin, A.N.; Hutchinson, R.A.; Buback, M.; Hesse, P. A novel approach for investigation of chain transfer events by pulsed laser polymerization. *Macromol. Chem. Phys.* **2011**, *212*, 699–707.

20. Barner-Kowollik, C.; Gunzler, F.; Junkers, T. Pushing the Limit: Pulsed Laser Polymerization of n-Butyl Acrylate at 500 Hz. *Macromolecules* **2008**, *41*, 8971–8973.

21. Lyons, R.A.; Hutovic, J.; Piton, M.C.; Christie, D.I.; Clay, P.A.; Manders, B.G.; Kable, S.H.; Gilbert, R.G. Pulsed-laser polymerization measurements of the propagation rate coefficient for butyl acrylate. *Macromolecules* **1996**, *29*, 1918–1927.

22. Beuermann, S.; Paquet, D.A.; McMinn, J.H.; Hutchinson, R.A. Determination of free-radical propagation rate coefficients of butyl, 2-ethylhexyl, and dodecyl acrylates by pulsed-laser polymerization. *Macromolecules* **1996**, *29*, 4206–4215.

23. Busch, M.; Wahl, A. The significance of transfer reactions in pulsed laser polymerization experiments. *Macromol. Theory Simul.* **1998**, *7*, 217–224.

24. Ahmad, N.M.; Heatley, F.; Lovell, P.A. Chain transfer to polymer in free-radical solution polymerization of n-butyl acrylate studied by NMR spectroscopy. *Macromolecules* **1998**, *31*, 2822–2827.

25. Peck, A.N.F.; Hutchinson, R.A. Secondary reactions in the high-temperature free radical polymerization of butyl acrylate. *Macromolecules* **2004**, *37*, 5944–5951.

26. Heatley, F.; Lovell, P.A.; Yamashita, T. Chain transfer to polymer in free-radical solution polymerization of 2-ethylhexyl acrylate studied by NMR spectroscopy. *Macromolecules* **2001**, *34*, 7636–7641.

27. Stickler, M.; Meyerhoff, G. The spontaneous thermal polymerization of mthyl-methacrylate: 5. Experimental-study and computer-simulation of the high conversion reaction at 130 °C. *Polymer* **1981**, *22*, 928–933.

28. Blavier, L.; Villermaux, J. Free-radical polymerization engineering: 2. Modeling of homogeneous polymerization of styrene in a batch reactor, influence of initiator. *Chem. Eng. Sci.* **1984**, *39*, 101–110.

29. Rier, T.; Srinivasan, S.; Soroush, M.; Kalfas, G.A.; Grady, M.C.; Rappe, A.M. Macroscopic mechanistic modeling and optimization of a self-initiated high-temperature polymerization reactor. In Proceedings of the 2011 American Control Conference, San Francisco, CA, USA, 29 June 2011–1 July 2011; pp. 3071–3076.

30. Asua, J.M.; Beuermann, S.; Buback, M.; Castignolles, P.; Charleux, B.; Gilbert, R.G.; Hutchinson, R.A.; Leiza, J.R.; Nikitin, A.N. Critically evaluated rate coefficients for free-radical polymerization, 5—Propagation rate coefficient for butyl acrylate. *Macromol. Chem. Phys.* **2004**, *205*, 2151–2160.

31. Nikitin, A.N.; Hutchinson, R.A. Effect of intramolecular transfer to polymer on stationary free radical polymerization of alkyl acrylates, 2—Improved consideration of termination. *Macromol. Theory Simul.* **2006**, *15*, 128–136.

32. Wang, W.; Nikitin, A.N.; Hutchinson, R.A. Consideration of macromonomer reactions in n-butyl acrylate free radical polymerization. *Macromol. Rapid Commun.* **2009**, *30*, 2022–2027.

33. Nikitin, A.N.; Hutchinson, R.A.; Kalfas, G.A.; Richards, J.R.; Bruni, C. The effect of intramolecular transfer to polymer on stationary free-radical polymerization of alkyl acrylates, 3-Consideration of solution polymerization up to high conversions. *Macromol. Theory Simul.* **2009**, *18*, 247–258.

34. Nikitin, A.N.; Hutchinson, R.A.; Wang, W.; Kalfas, G.A.; Richards, J.R.; Bruni, C. Effect of intramolecular transfer to polymer on stationary free-radical polymerization of alkyl acrylates, 5-Consideration of solution polymerization up to high temperatures. *Macromol. React. Eng.* **2010**, *4*, 691–706.

35. Yu, X. Kinetic Modeling of Acrylate Polymerization at High Temperature. Ph.D. Thesis, Northwestern University, Evanston, IL, USA, 2008.

36. Kruse, T.M.; Woo, O.S.; Broadbelt, L.J. Detailed mechanistic modeling of polymer degradation: Application to polystyrene. *Chem. Eng. Sci.* **2001**, *56*, 971–979.

37. Zhu, S.P. Modeling of molecular weight development in atom transfer radical polymerization. *Macromol. Theory Simul.* **1999**, *8*, 29–37.

38. Villermaux, J.; Blavier, L. Free-radical polymerization engineering. 1. A new method for modeling free-radical homogeneous polymerization reactions. *Chem. Eng. Sci.* **1984**, *39*, 87–99.

39. Kim, D.M.; Iedema, P.D. Modeling of branching density and branching distribution in low-density polyethylene polymerization. *Chem. Eng. Sci.* **2008**, *63*, 2035–2046.

40. Benyahia, B.; Latifi, M.A.; Fonteix, C.; Pla, F.; Nacef, S. Emulsion copolymerization of styrene and butyl acrylate in the presence of a chain transfer agent. Part 1: Modelling and experimentation of batch and fedbatch processes. *Chem. Eng. Sci.* **2010**, *65*, 850–869.

41. Martinez, E.C. Model discrimination and selection in evolutionary optimization of batch processes with tendency models. *Comput. Aided Chem. Eng.* **2005**, *20*, 463–468.

42. Fotopoulos, J.; Georgakis, C.; Stenger, H.G. Use of tendency models and their uncertainty in the design of state estimators for batch reactors. *Chem. Eng. Process.* **1998**, *37*, 545–558.

43. Fotopoulos, J.; Georgakis, C.; Stenger, H.G. Effect of process-model mismatch on the optimization of the catalytic epoxidation of oleic acid using tendency models. *Chem. Eng. Sci.* **1996**, *51*, 1899–1908.

44. Filippi, C.; Greffe, J.L.; Bordet, J.; Villermaux, J.; Barnay, J.L.; Bonte, P.; Georgakis, C. Tendency modeling of semibatch reactors for optimization and control. *Chem. Eng. Sci.* **1986**, *41*, 913–920.

45. Martinez, E.C.; Cristaldi, M.D.; Grau, R.J. Design of dynamic experiments in modeling for optimization of batch processes. *Ind. Eng. Chem. Res.* **2009**, *48*, 3453–3465.

46. Mueller, P.A.; Richards, J.R.; Congalidis, J.P. Polymerization reactor modeling in industry. *Macromol. React. Eng.* **2011**, *5*, 261–277.

47. Maeder, S.; Gilbert, R.G. Measurement of transfer constant for butyl acrylate free-radical polymerization. *Macromolecules* **1998**, *31*, 4410–4418.

48. Moad, G.; Solomon, D.H. *The Chemistry of Free Radical Polymerization*; Elsevier: Oxford, UK, 1995.

49. Derboven, P.; D'hooge, D.R.; Reyniers, M.-F.; Marin, G.B.; Barner-Kowollik, C. The long and the short of radical polymerization. *Macromolecules* **2015**, *48*, 492–501.

50. Barth, J.; Buback, M.; Hesse, P.; Sergeeva, T. Termination and transfer kinetics of butyl acrylate radical polymerization studies via SP-PLP-EPR. *Macromolecules* **2010**, *43*, 4023–4031.

51. Hulburt, H.M.; Katz, S. Some problems in particle technology. *Chem. Eng. Sci.* **1964**, *19*, 555–574.